Mike West
2014

FENCE
NOV 1994
KENTIGERN HOUSE
LIBRARY, GLASGOW

Ministry of Defence
KENTIGERN HOUSE LIBRARY SERVICES, RM 1212
STREET, GLASGOW G2 8EX

Trident facilities

Proceedings of the conference organized by the Institution of Civil Engineers, held in London on 14 April 1994

Edited by F. D. R. Yell

Conference organized by the Institution of Civil Engineers.

Organizing committee:
F. D. R. Yell, Associated British Ports, (Chairman);
K. Partington, TBV Consult Ltd;
F. C. Ince, Associated British Ports, Barrow & Silloth;
J. Wotton, Crouch Hogg Waterman;
R. F. Hodge, Gardiner & Theobald Management Services;
J. G. Gittins, Stothert & Pitt.

The Institution of Civil Engineers and the Authors of papers presented at this conference acknowledge the support given by the Ministry of Defence.

First published 1994.

Distributors for Thomas Telford books are
USA: American Society of Civil Engineers, Publications Sales Department, 345 East 47th Street, New York, NY 10017-2398
Japan: Maruzen Co. Ltd, Book Department, 3–10 Nihonbashi 2-chome, Chuo-ku, Tokyo 103
Australia: DA Books and Journals, 648 Whitehorse Road, Mitcham 3132, Victoria

Classification
Availability: unrestricted
Content: collected papers
Status: refereed
User: maritime and military engineers

ISBN 0-7277-1992-0

A CIP catalogue record for this book is available from the British Library.

© Authors and the Institution of Civil Engineers, 1994, unless otherwise stated.

Papers or other contributions and the statements made or the opinions expressed therein are published on the understanding that the author of the contribution is solely responsible for the opinions expressed in it and that its publication does not necessarily imply that such statements and/or opinions are or reflect the views or opinions of the organizers or publishers.

All rights, including translation, reserved. Except for fair copying, no part of this publication may be reproduced, stored in a retrieval system or transmitted in any form or by any means electronic, mechanical, photocopying, recording or otherwise, without the prior written permission of the Publications Manager, Publications Division, Thomas Telford Services Ltd, Thomas Telford House, 1 Heron Quay, London E14 4JD.

Published on behalf of the Institution of Civil Engineers by Thomas Telford Services Ltd, Thomas Telford House, 1 Heron Quay, London E14 4JD.

Printed and bound in Great Britain by The Cromwell Press, Melksham, Wilts.

Contents

A general review of requirements and their project management.
R. F. HODGE   1

Design and construction of the new dock entrance. F. C. INCE and
P. D. H. HATLEY   15

Disposal of dredged spoil and channel design. Captain J. W. GREEN   30

Dredging for Trident class submarines. R. J. GARDNER and D. H. ROBERTS   44

Coulport and Faslane general management. K. PARTINGTON   57

Safety case development for Trident facilities. B. V. DAY and
P. H. DAVIDSON   69

The Faslane reclamation. J. H. WOTTON and D. K. BELSHAM   82

Design and construction of the Shiplift and Finger Jetty. W. J. PATERSON,
R. DUNSIRE and D. A. BLACKBURN   96

The Faslane Syncrolift. G. A. STOKOE and J. R. BERRY   110

Seismically qualified jib cranes. N. P. HANCOCK   124

HV distribution and jetty support services. R. BALL,
EurIng A. D. CAMERON and H. A. CHERRY   138

The floating explosives handling jetty. J. R. WARMINGTON, J. P. ABELL
and M. W. PINKNEY   153

Design and construction of the jetty access roads and support area.
B. T. THOMAS and E. R. SHARPLES   167

A general review of requirements and their project management

R. F. HODGE, Regional Director, Gardiner & Theobald Management Services Ltd

SYNOPSIS

1   In the late 1970s, VSEL decided to modernise its shipyard at Barrow in Furness and between 1982 and early 1992 it caused the building of the finest warship and submarine yard in the UK and perhaps the world, together with associated works in the dock system, at the dock entrance and to the channel to the open sea.

This paper briefly describes the Shipbuilding Works Modernisation (which has been the subject of an earlier paper given to the Institution) and reviews more particularly the subsequent works needed to enable the transits of Vanguard Class Submarines between the dock and the open sea.

BACKGROUND

2   The contract for the first of the four UK Trident armed submarines (HMS Vanguard) was placed by the MOD with VSEL in April 1986. The contract was the largest for a single warship ever placed in the UK. The contracts for the remaining three Vanguard Class submarines have been placed since and they are in various stages of construction.

SHIPBUILDING WORKS MODERNISATION

Generally:

3   The Works were modernised between 1982 and 1988. The complex is unique in the manner in which it enables vessels to be built. Hull steelwork is prepared in the modernised existing works, transported to a new construction facility, where the vessels are built in a controlled environment in an enclosed building. On completion they are rolled out on bogies to a shiplift and lowered into the water.

REQUIREMENTS AND PROJECT MANAGEMENT

4   The work may be classified under the following headings:

   1 - Site Creation
   2 - Quays
   3 - Shiplift
   4 - Construction Facility

Site Creation:

5   No area was available within the boundaries of existing works on Barrow Island for the size of the site required. However, the problem was solved by reclaiming part of a large dock (Devonshire Dock) adjacent to the existing works. In 1982 a contract was let to Balfour Beatty Construction/Holland Dredging Joint Venture to fill the major portion of the dock with sand. The sand, obtained from Roosecote Sands in Morecambe Bay, was removed by a floating dredger and hydraulically pumped via a floating and land based pipeline some four kilometres to the dock in the centre of Barrow. In a period of 8½ months 2,680,000 tonnes of sand were pumped, providing an additional site area of 7½ hectares. The sand was subsequently compacted by vibro-compaction and 'surcharging'.

Quays:

6   Reinforced concrete quays were built through the filling. They provide two servicing levels founded on a combination of diaphragm walling and large bored piling to a depth of 42 metres and are among the deepest such construction in Europe. Upon completion of the quays, surplus sand was removed by a combination of land and marine dredging. Quays were executed by Alfred McAlpine Construction Limited with Cementation Piling & Foundations Limited as a major domestic sub-contractor.

Shiplift:

7   The Contract for the shiplift was placed with NEI Clarke-Chapman Limited. It is 162 m long x 22 m wide and has maximum lifting capacity of 24,000 tonnes, through 108 winches. Vessels are lowered into the dock water at a rate of 200 mm/minute.

Construction Facility:

8   The most obvious part of the whole project is the Construction Hall, the contract for which was awarded in July 1984 to Alfred McAlpine Construction Limited.

The Hall is 269 m long, 67 m wide and 50 m high and can contain up to five vessels at the same time. Components for the outfitting of the vessels are fed into the Hall from 14,000 square metres of workshop on the sides of the building. The workshops are separated by amenity towers for the workforce and contain toilets, lockers, mess rooms, etc. Vessels are built on cradles on the floor of the Hall. The floor incorporates a rail system and the vessel in whole or in parts can be moved between these lines and out onto the Shiplift.

Contract Details etc:

9   The works were procured using the ICE 5th Edition with Special Conditions, the !Employer' being VSEL and the 'Engineer' being R T James & Partners.

10 In October 1988, R T James & Partners received High Commendation in the British Construction Industry Awards (Civil Engineering). Gardiner & Theobald were the quantity surveyors.

TRANSITS OF VANGUARD CLASS SUBMARINES

Generally:

11 This Review deals with works needed to enable the transits of Vanguard Class submarines between Barrow Docks and the open sea and to increase the frequency for transits. Each transit comprises:

    a)   a level dead tow through the dock system;

    b)   a stop over at Ramsden Dock Basin;

    c)   a self-propelled exit or entry down the channel with tugs in attendance and assisting at the turn into the Entrance.

The Requirements:

12 The works required to prepare the route (and the various techniques) for the transits and to increase the number of occasions on which exit or entry transits are possible comprised:

    a)   various works in widening the existing dock fairway;

    b)   upgrading the facilities at Ramsden Basin for a stop over;

REQUIREMENTS AND PROJECT MANAGEMENT

- c) a new dock entrance and gate with a deeper sill;
- d) designing a channel to suit the lowered sill and the characteristics of the submarine and the accompanying tugs;
- e) dredging the channel and installing new and improved navigation aids, tide gauges, etc;
- f) associated work and advice.

The Principal Parties:

13   The principal parties having an input into effecting the works were:

   A   The Ministry of Defence

   B   VSEL

   C   Associated British Ports (who own and operate the port of Barrow)

Principal Contractual Arrangements:

(MOD/VSEL)

14   A contract for the whole of the works was placed by the Ministry of Defence with VSEL in October 1988 and was known as the Contract for the Exit and Entry Services at Barrow for Vanguard Class Submarines. The contract was:

- a) based on Stores Purchases Form GC/Stores 1, Edition April 1979 with Special Conditions
- b) to be completed by December 1991, ie in time for the transit of HMS Vanguard early in 1992.

(VSEL/ABP)

15   VSEL and ABP entered a Special Contract which required ABP to act as VSEL's agent and give particular support as design and supervising Engineers, particularly in respect of civils works.

(VSEL/Others and ABP/Others)

16   Every effort was made to make each contract a fixed price lump sum contract. Contracts of a civil engineering nature were based on the ICE 5th Edition with Special Conditions. The Special Conditions included the deletion of "ground condition" and

"valuation by admeasurement (remeasure)" clauses.

Project Management:

(Objectives)

17   The objective of the project management was to design and procure to the satisfaction of the Client particular works to programme and budget, whether in the 'lead' contract or any sub-contract thereafter. The activities to achieve the objective can be sub-divided into 'pre-contract' and 'post contract', ie:

   A)   Pre-contract -

   i)   deciding on an appropriate package and affixing the budget and programme.

   The ideal was regarded as a fixed price lump sum.

   ii)   defining requirements with design completed as far as is practicable in the timescale available.

   iii)   calling tenders for the work based on a form of contract and conditions appropriate to the stage of design development.

   iv)   evaluating the tenders and placing the contract with the successful tenderer.

   B)   Post contract -

   i)   supervision and monitoring contracts individually and collectively in respect of:

   a)   construction;
   b)   progress;
   c)   programme;
   d)   contract sum (as/if affected by variation) compared with budget.

   ii)   in the event that the above monitoring identified that the objective was not being achieved, identification of appropriate options to remedy the situation and implementation of any agreed remedial action.

REQUIREMENTS AND PROJECT MANAGEMENT

(Generally)

18   The lead contract (MOD/VSEL) required that each party set up its own project management team with VSEL's team managing the works.

The author was seconded to VSEL to act as Project Manager. VSEL's Management Team reported to MOD Project Manager under the following standard headings/requirements:

A)   Communication and interface -

- with the Client, ie MOD

- between all works within the contract

- between all levels of the Vanguard Submarine Project

B)   Progress -

- providing a monitoring system

C)   Programme -

- generating and operating a network and bar charts

D)   Budget -

- preparation, updating and monitoring against progress and programme

E)   Design and construction -

- design and specification
- Quality Plan and Control

(The First Tasks)

19   The first tasks of the Management Team were to:

A)   analyse the requirements into work sub-contracted and work to be executed 'in house'. The requirements were analysed under the following headings:

   a)   Civils and building
   b)   The channel
   c)   Shipbuilder items

B) prepare a programme initially based on a simple bar chart, each bar representing a proposed sub-contract or shipbuilder item.

C) prepare a detailed budget, each item of the budget again being a proposed sub-contract or shipbuilder item.

(The Civils and Building Package)

20 The Civils and Building package was analysed as indicated in Figure No 1. The package was increased during the project to include:

A) a pumped water dock impounding scheme;

B) a lead in system at the lock and at Michaelson Passage.

21 The civil works in respect of the new gate and the structure to take it (ie the pneumatic caisson) will be described in Paper 2 by Mr Frank Ince, ABP's Port Engineer. The works were available in time for the exit of HMS Triumph on 6 November 1991.

(The Channel Package)

22 The Channel Package was analysed as indicated in Figure No 2.

23 The Channel design and the associated dredging will be described in Papers 3 and 4 by Capt John Green, ABP's Port Manager and Harbour Master, and by Mr R Gardner of the Barrow Channel Dredging Joint Venture.

(The Shipbuilder Package)

24 The Shipbuilder Package included:

a) the development of a plan for shore support during the Walney Channel transit including a computer based speed and distance monitoring and prediction system.

Note: Great care was taken during the works to ensure that the various measuring systems/ predictions by the several contractors involved were compatible with the system, eg:

i) horizontal fixing system (Microfix) shore stations were used in all dredging, surveying and speed and distance monitoring.

REQUIREMENTS AND PROJECT MANAGEMENT

- ii) the wave data used was that from the three new tide gauges and was interpreted (reduced) in the same fashion.

- iii) levels related to Ordnance Datum were affixed at the entrance and on fixed beacon structures in the channel thus minimising errors in determining sea bed levels.

b) Advice on what (if any) emergency facilities should be provided in the event that transit through the channel should be impeded in any way.

c) Development of a towing technique including:

- numbers and size;
- attachment;
- disposition;
- special temporary hardware including 'quick release' mechanisms

d) Determination of the requirements for changes to ship condition for the exit.

e) Provision of nuclear safety justification.

f) Definition of support services and facilities required in Devonshire Dock Basin.

g) Design and procurement of catamarans and fendering.

h) Obtaining of nuclear berth classification.

(Communication and Interface)

25 The MOD chaired monthly Progress Meetings at which VSEL reported under standard headings and agreed remedial actions (if any) where progress, programme or budget were endangered.

26 VSEL chaired monthly Progress Meetings with its 'in house' departments and its sub-contractors and with ABP; it also attended meetings of ABP's sub-contractors. Throughout the meetings, a common agenda was used, ie:

- a) Progress
- b) Programme
- c) Budget/Contract Sum
- d) Finance/Expenditure
- e) Remedial actions (if necessary)

27   The information was continually collated by VSEL into a monthly report to MOD who were thus able to act as a fully informed Client and agree priorities and remedial actions as the work proceeded for individual parts of the project and for the project as a whole.

(Progress)

28   The VSEL Project Manager and each major Contractor 'down the chain' was obliged to submit written reports on a monthly basis to its Employer/Engineer comparing predicted with actual progress; the validity of each report was examined at the Progress Meeting.

(Programme)

29   VSEL's initial bar chart was drawn up with the following principles:

   A)   that as much of the works as possible should be completed as soon as practical to avoid 'log jams' towards the end of the project.

   B)   dredging should be completed as late as possible to minimise subsequent maintenance dredging associated with the transit of Vanguard.

30   This led to a target of completing all civils work in the first two years of the project and the dredging and navigation aids etc in the third year.

31   It was further recognised that the improved predictions and surveying of the extended channel were dependant on the tide gauges being installed as soon as practicable.

32   VSEL progressively developed the bar chart into a critical path network as the details and interdependencies of the Project Team's programmes and each Contractor's programme became known.

33   As far as was practical, each bar or activity represented a financial package. The effect of actual against predicted progress was constantly reviewed overall and individual 'floats' determined (to assist in decisions on remedial actions required if programme/budget were jeopardised).

REQUIREMENTS AND PROJECT MANAGEMENT

(Budget)

34   In a similar manner to Progress/Programme, the VSEL original budget was constantly reviewed to compare actual and budgetted expenditure. Payments against each contract were analysed in detail and compared with the contractor's applications. Discrepancies (claims) between the two were identified and a judgment made as to ultimate validity. Appropriate allowances for claims and variations were then added to or deducted from the individual contract budget on a monthly basis. The results were then collated in a project budget at no more than quarterly intervals.

35   Throughout the various contractual arrangements VSEL did not delegate the authority to vary the works if that variation affected programme or contract sum.

36   Where remedial actions became necessary, each possible action was separately assessed in value and programme effect; the optimum action was then agreed with the MOD before implementation.

(Design and Construction)

Generally:

37   Design was achieved by VSEL by -

   1   'in house' expertise

   2   employing Civil Engineering Consultants

   3   causing 'design and construct' contracts to be placed for civil works

   4   employing specialist advisors (eg channel designers)

Quality Control:

38   VSEL was required to provide MOD with a Quality Plan for the whole project. Basically VSEL's own activities were subject to a quality plan which included AQAP.13 (NATO Software Quality Control System Requirements).

39   Appropriate equivalent requirements were included in every Contract thereafter in the chain and audits were carried out intermittently.

Summary

(Programme)

40 The MOD/VSEL contractual completion date of the three year project was end of December 1991. It was extended by mutual agreement to March 1992 for reasons outwith the Contract.

(Budget)

41 The initial budget increased during the project. However, the MOD was advised considerably in advance of the cost of variations and the end cost was within the predicted budget.

(Quality)

42 At all contract levels of the project, appropriate certification of quality was ensured and the MOD was able to accept the project as complete to its satisfaction in March 1992.

Conclusion

43 On 22 October 1992, HMS Vanguard, made the first transit from Devonshire Dock through the new entrance and out to sea via the purpose designed and dredged channel for sea trials. It has made its entry and final exit transits since. The project known as Exit and Entry Services for Vanguard Class Submarines met the Client's requirements.

REQUIREMENTS AND PROJECT MANAGEMENT

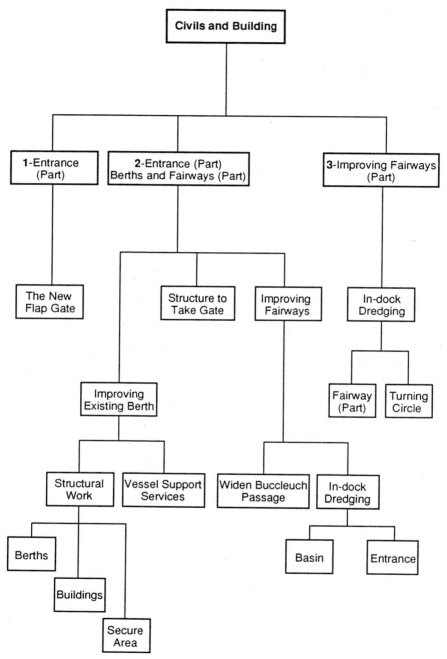

**Figure 1
Barrow Exit and Entry Services
CIVIL AND BUILDING PACKAGE**

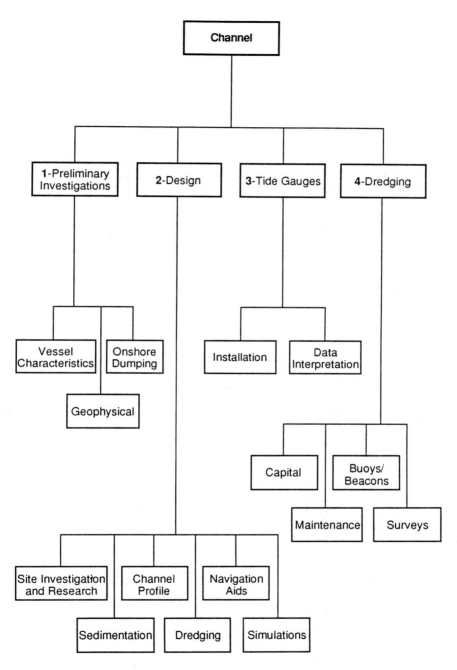

**Figure 2
Barrow Exit and Entry Services
THE CHANNEL PACKAGE**

REQUIREMENTS AND PROJECT MANAGEMENT

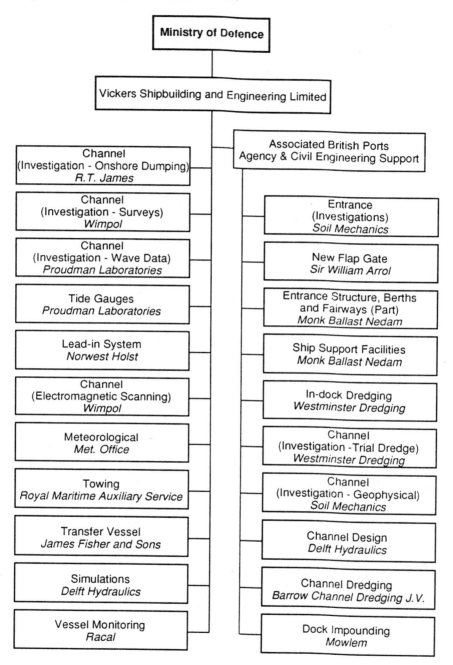

Figure 3
Barrow Exit and Entry Services
CONTRACTUAL ARRANGEMENTS (EXCLUDING IN-HOUSE WORK)

Design and construction of the new dock entrance

F. C. INCE, Port Engineer, Barrow, and P. D. H. HATLEY, Engineering Assistant, Hull, Associated British Ports

SYNOPSIS. The new dock entrance at Barrow was completed in February 1992. The most interesting aspects are the use of a bottom hinged flap dock gate and the employment of the pneumatic caisson technique in the construction of the reinforced concrete entrance structure. The 50mx50mx24m structure was constructed above ground level and sunk 22m into the final position using high pressure water jetting. Compressed air was used to keep the excavation chamber dry.

INTRODUCTION
1. The new dock entrance at Barrow-in-Furness was constructed between July 1989 and February 1992 as part of the works carried out by port owners Associated British Ports (ABP) to meet the requirements of Vickers Shipbuilding & Engineering Ltd. (VSEL) to facilitate the transits of Vanguard Class submarines between the Port of Barrow and the open sea. The works were part of the Civils and Building package referred to in Paper 1. The new entrance is 36.6m wide and will enable Barrow to handle vessels of up to 11.5 metres draught on spring tides.

Project Management
2. ABP's contract with VSEL for the provision of exit and entry services at Barrow included the following works:
New Dock Entrance - civil works to form the new entrance and improve an existing berth
                   - dock gate
                   - support facilities in Ramsden Dock Basin
Deep Dredged Channel - channel design
                     - dredging
                     - provision of navigation aids
Miscellaneous Facilities - widening of Buccleuch Passage
                         - dock impounding system
                         - in-dock dredging
The scope of this paper has been limited to the new dock gate and civil works required to form the new dock entrance.
3. ABP were required to set up a project management team reporting to VSEL's Project Manager and responsible for the administration, supervision, and monitoring/reporting of

progress and programme and budgetary control in respect of all the works encompassed in the Contract with VSEL. Quality Assurance in accordance with AQAP1 was a requirement both for ABP and ABP's contractors.

DOCK GATE

4. The type of gate chosen for the new entrance is a bottom hinged flap gate (Fig.1). This type of gate should avoid the inevitable leakage which develops over time with traditional mitre gates. Other advantages are that both the gate structure and accompanying civil works are relatively simple to construct and the gate can be made buoyant for stepping and removal. Disadvantages are that a recess is required below cill level and the gate cannot easily accept a reverse head, can only open against a small head and cannot be maintained in-situ. A further disadvantage compared with mitre gates is the difficulty in arranging make-shift closing devices in the event of failure of the operating equipment. For this reason, particular attention was paid to the provision of duplicate equipment.

5. In February 1989, a contract was awarded to Sir William Arrol for the design and construction of the gate. As shown in Fig. 1, the gate structure was designed with upper and lower watertight box sections to provide buoyancy and rigidity. The upper box also comprises the main load carrying member spanning 37m across the entrance when water is retained. The main meeting faces of the gate comprise steel billets at the upper box section and rubber billets elsewhere. Music note section rubber seals are used throughout.

6. The Contractor's design for the operation of the gate made use of a single fall of rope led from a hydraulic winch at one side of the entrance, through a series of pulleys, to the other end of the gate and anchored. This method of reeving ensures simultaneous hauling of each end of the gate. This system was upgraded by the provision of a winch at both ends of the rope, either one of which can be used to operate the gate. The hydraulic system at each side, which also power locking bolts and push-off rams, is driven by either electric or diesel pumps.

NEW ENTRANCE STRUCTURE
The Problem

7. Following award of the gate contract in February 1989, tenders were invited for the design and construction of the civil engineering works based upon a simple general arrangement drawing indicating the required position and dimensions (Fig.2).

8. Two main areas of difficulty were envisaged.
   i. It was essential that operation of the existing entrance was maintained. Due to the proximity of the new entrance, the conventional method of constructing the structure within an excavated cofferdam was expected to

raise questions of stability of the existing entrance walls with vibration and ground movement anticipated problem areas.

ii. Site investigations had revealed that further possible problems associated with excavation would be ingress of water and stability of the excavated base due to the permeability of the ground strata. An extensive ground water control system was proposed by the site investigation report but the practicality of this, and possible effect on adjacent structures, was questionable.

The solution

9. The tenders submitted proved to be an interesting example of design and construct bidding. A wide range of methods of construction were proposed and this posed the difficult problem of selecting the most acceptable both financially and technically. Various proposals based on the cofferdam technique were submitted but technical and financial reservations cast doubt over these.

10. The chosen solution submitted by a Monk/Ballast Nedam Joint Venture, was not in any way envisaged by ABP/VSEL. The Contractor proposed to construct a 50m x 50m x 24m high reinforced concrete structure above ground level and subsequently sink this 58,000t structure 22m to the correct position using high pressure water jetting to remove the ground strata (Fig.3). Compressed air would be used to keep the excavation chamber dry and so the entrance structure itself would become a pneumatic caisson.

11. Although there is extensive experience of this use of the pneumatic caisson technique in Europe, the proposed structure would be the largest example outside of Japan (Fig.3).

Excavated spoil would be removed from the 2m high working chamber by pumping and after completion of sinking, this chamber would be filled with concrete to provide additional factor of safety against uplift (Fig.4).

Before the contract was placed, the tenderer was asked to satisfy ABP and VSEL that the proposed method of construction was feasible in any circumstances which were likely to arise. Although sand was the preferred soil strata for the sinking process, other materials likely to arise, such as gravel or clay, could be dealt with although with a consequent slowing down of progress. Confirmation was also given that obstructions could be dealt with, the general methodology being a choice of -

i) breaking up into fragments which could be removed through the access chambers,
ii) removing sections as sinking proceeded e.g. piling, or
iii) removing the ground around allowing the obstruction to be taken down with the caisson.

12. A further interesting area of design in the successful tender was the use of the Mannesmann combi-wall technique to construct the retaining walls which comprised the roundheads

adjacent to the caisson structure. This form of construction was also used to strengthen and deepen the existing gravity wall along the south side of Ramsden Dock Basin.

Construction

13. The proposed construction sequence is shown in Fig.5 a-e. As can be seen from Fig.5b, extensive temporary works were required in order to produce a suitable surface upon which to construct the caisson. The main objectives were to
1. remove existing features, such as the dock wall and slope protection,
2. remove the top 2m or so of the existing ground and replace with uniform sand, and
3. extend the working area into both the dock basin and Walney Channel by the use of sheet-piled retaining walls and sand fill (Fig.6).

In order to ensure even sinking of the caisson during the critical initial 6 to 10m, uniform ground material was required. This was achieved by replacing the top 2m of the existing ground with uniform sand and using a vibro-pile to ensure uniform compaction of this material, the sand infill and the existing ground. This was checked using Dutch Cone Penetration tests.

14. Construction of the permanent external roundheads, using 1220mm and 1016mm dia. tubes each linked with two Larrsen 16W or 12W sheet piles, was also included in this stage. Tie rods of up to 150mm dia. were employed at several levels. To ensure the continued integrity of the existing dock entrance ICE vibrating hammers were mainly used for the installation of piles. Driving of the combi-wall tubes was carried out using a purpose made frame placed at ground level to ensure correct spacing for subsequent installation of the Larrsen sheets. Each pile was manoeuvred until it was seen to be vertical using two theodolites. Although this method of installation was quick and flexible, the tolerance on plumbness required for installation of the infill sheets was high and some problems arose. Difficulties in driving the infill sheets, due to the light section employed, were experienced in some areas and were overcome using a purpose-made jetting device which was fabricated to fit the piles clutches.

15. Construction commenced in July 1989 and the complex temporary works comprising Stage I continued until March 1990, overlapping slightly with the caisson construction. The first step in constructing the caisson was the formation of a cutting edge around the perimeter. This comprised a precast concrete outer face 2m deep incorporating a steel channel at the lower edge and in-situ concrete sloping inner face. Construction of the caisson was a sizeable project in itself. A total of 23,000 cu.m of concrete was required together with a large quantity of reinforcement, a high proportion of which was 40mm diameter. In order to ensure durable concrete and to minimise thermal problems in the base which is up to 9m deep, grade 40 concrete containing 70% G.G.B.S. cement replacement was used.

Placing was by pump and a super-plasticiser was used to achieve a target slump of 90mm for the structural concrete and 150mm for the mass concrete used to fill the void beneath the caisson on completion of sinking.

16. Single pours of up to 1000 cu.m were used to construct the base with pours of up to 400 cu.m being used for the walls. In order to minimise stresses due to settlement, it was necessary for the caisson to be constructed in a pre-determined sequence, the thinner central section being completed last. The critical loading cases for the design of the caisson structure were generally those arising during the sinking process due to possible uneven support of the cutting edges. In addition to the structural reinforcement there were many cast-in items required for the sinking process such as pipes for water distribution to the jets, excavated material pumped out and compressed air. Cable ducts and a water main across the entrance were also accommodated.

17. Construction of the caisson ready for sinking was completed just before Christmas 1990 (Fig.7) and during the latter stages German sub-contractor Wayss and Freytag carried out installation of their specialist equipment ready for the sinking process to begin. The U-shaped structure through which shipping will pass after completion can be seen in Fig.7. In order to prevent the ingress of soil and water during the sinking process, a steel bulkhead was constructed across each end. These consisted of interlocking sheet pile sections supported by walings and raking struts secured to the base of the caisson(Fig.8). Although the erection of the inner (dock basin) bulkhead was well advanced at completion of the concrete structure, the outer bulkhead was largely constructed after sinking had commenced.

Sinking

18. The operation of sinking the caisson was started by forming the initial working chamber by hand using one of the four access holes through the caisson base slab to gain entry and for the removal of spoil. Work at this time was carried out at atmospheric pressure and coincided with the erection of 24m high access shafts, blister locks and the main decompression chamber to the other three access holes (Fig.9). Work was also carried out to install the three main suction pumps, located on the floor of the caisson, into waterproof chambers manufactured from 2m diameter steel tube. There were two reasons for having the access to the working chamber at the top of the caisson rather than on the caisson floor and for having the pumps sealed. The first was as a safety precaution against flooding following possible damage to the bulkheads and the second as a provision against the caisson being too buoyant to sink despite the provision of a bentonite slurry reducing friction on the outside of the caisson. The specific gravity of the caisson prior to flooding would come very close to 1.0 as sinking progressed and it was considered necessary to allow for the flooding of the caisson to provide kentledge if

problems arose. The process of sinking involves removing sufficient material from under the centre of the caisson that its dead weight is sufficient to overcome the friction against the sides and produce a shear failure in the ground beneath it. This shear failure occurs at the cutting edges and their profile forces the material toward the centre of the caisson where it is removed allowing the process to continue.

19. Once the working chamber had been formed and the water jetting equipment fixed to the supply pipework cast into the caisson base slab, the fourth access hole was sealed and the working chamber pressurised. Sea water was used for jetting away the ground beneath the caisson and was pumped to the jets from pontoon mounted pumps in the dock basin thereby allowing jetting to continue at all states of the tide. The jetted material was carried in suspension to the suction pumps which removed it to a settling pond constructed at the south side of the caisson.

20. The working chamber had to be pressurised in order to prevent the ingress of water through the highly permeable sands and gravels below the caisson. Compressed air working does however require stringent safety procedures to be used and the system of work had to be approved by the Health and Safety Executive (HSE). The method of 'decanting', which involves the rapid decompression of workers who then walk to a main decompression chamber for recompression and slow decompression, was not permitted by the HSE. This meant that the main decompression chamber had to be located at the exit from the working chamber.

21. Particular attention was paid to the medical supervision of the personnel working in compressed air conditions. All personnel were subject to an extensive preliminary examination and regular check-ups and medical lock attendants were responsible for monitoring and recording all movements in and out of the chamber. Another important safety measure included the provision of an on-site medical lock where any worker suffering with decompression problems could be treated. Decompression times were calculated according to the Blackpool Tables and then increased further to combat the problems of workers being either wet or cold during decompression and having a lower rate of circulation. Subject to passing the medical examinations, it was found possible to employ local labour to carry out much of the compressed air work.

22. During the sinking of the caisson, a number of obstructions were encountered (Fig.10). These included timber piles and the remains of the concrete foundations of the old dock wall. The timber piles were simply cut off in lengths as the caisson sank but the concrete proved to be a more difficult problem. Initial attempts to remove it with pneumatic breakers proved time consuming and blasting was therefore used to speed up fragmentation. The resulting rubble was then transported by hand across the working chamber and loaded into buckets which were hoisted up one of the access shafts and discharged

through a blister lock.

23. Most of the ground beneath the caisson comprised sands and gravels which were readily removed by the jetting process. However removal of the layers of boulder clay which underlay the sands and gravels proved to be a more time consuming operation. The jetting method of excavation used is ideally suited to non-cohesive materials which constitute most of the excavated material, however the clays proved much more difficult to liquify and transport to the suction pumps. Ideally excavation by mechanical means would have been more suitable on the cohesive material but it was not considered necessary to provide an additional excavation method for the relatively small quantity of clay involved. The choice of excavation method was based on the original site investigation and emphasises the importance of this part of the work.

24. One benefit of the clay is that the air pressure was limited to 2.2 bar in the working chamber due to its lower permeability. It had originally been anticipated that 3.0 bar would be required toward the end of sinking and consequently there was a reduction in the anticipated decompression times required. Despite this each shift still had to work for 4 hours and decompress for 4 hours resulting in only 12 hours of productive work from the three shifts in a 24 hour period.

25. Controlling the position of the caisson during sinking was an important part of the operation and there were two main problems to deal with.

1. Limiting torsion of the caisson was very important as excessive rotation of the walls in opposite directions would have caused the caisson to break its back. This aspect was monitored using 2 inclinometers positioned on the top of the walls in conjunction with the techeometrical surveying mentioned below.

2. The final position of the caisson had to be achieved within a 150mm tolerance in all directions. All final lines and levels were achieved by the use of secondary concrete pours to areas such as the wall tops and the gate jambs. As these pours are only able to accommodate 150mm of error in position it was essential that the sinking tolerances were met.

26. Position control of the caisson during sinking was achieved by continual accurate techeometrical surveying combined with selective excavation. Direction could only be slightly adjusted by lowering one side of the caisson whilst torsion was limited by ensuring that no corner was significantly higher than the others. At the end of sinking the caisson achieved its final plan position to an accuracy better than 100mm and level better than 50mm.

27. Once sinking had been completed the working chamber had to be filled with concrete to reduce settlement after flooding and to achieve a factor of safety of 1.3 against buoyancy for the unflooded condition. Reinforcement cast into the underside of the base slab was bent down into the working chamber and

super-plasticised concrete placed via the service pipe work used for jetting. Super-plasticised concrete was used as it is self compacting and traditional compaction methods would have been impossible in this situation. The working chamber at this time was still pressurised and so concrete pumps were used to place the concrete against the pressure. A total of 4500 m³ were placed over a period of 3.5 days.
Finishing Work
28. Returning to the construction sequence described in Fig.5b, it can be seen that item 1 "Building and sinking of Caisson" had a total duration of 13 months. It was then necessary to complete the remaining items of work described in Fig.5b and c before the new dock entrance could be opened to shipping. In general terms, these remaining works comprised
a) removal of temporary works and fill,
b) dredging a new outer channel and deepening the dock basin,
c) completion of construction of the roundheads, aprons and buildings, and
d) installation of the new flap gate.
29. Completion of the works around the caisson within the shortest possible time proved to be a very complex operation because of the inter-dependence of several items. For example, the construction of a machinery house at either side of the entrance was required in time to accommodate the gate machinery prior to commissioning. In order to meet this programme, the north machinery house, which was also to form part of the new two-storey control building, was constructed prior to completion of sinking of the caisson using precast concrete units. These units were supported on two foundation slabs, one above the other, with in-built jacking points to facilitate correction for the 'draw-down' effect of the caisson. Fig.11 shows progress at the completion of sinking (April 1991) and Fig.12 as in August 1991. The south machinery house was supported partly on the south roundhead wall and partly on bearing piles. Subsequent to construction of this machinery house, excavation of the south roundhead was required to facilitate installation of the tie rods (Fig.13).
30. In addition to the above, further finishing works were required to the caisson structure itself because of the tolerance associated with the sinking process. These works comprised critical areas of concrete and all the gate 'cast-in' items. Areas of concrete for which the position or level was critical included the gate cill and jambs, the top 0.5m of wall on either side of the structure, and concrete in the immediate vicinity of each of the gate items. Construction of much of this 'secondary concrete' was required to extremely close tolerances in order to ensure correct functioning of the gate. The meeting faces for the gate, comprising the cill 37m long and two vertical jambs each 15m in length, were constructed within an overall tolerance of ± 3mm. This was achieved by the use of precast concrete units 9m long mounted on adjustable

anchorages. Cast-in steel meeting faces 3.7m long, supplied by the gate contractor, were used in the upper portion of each jamb where the main horizontal thrust from the gate would be imparted. A tolerance of ± 0.5mm was achieved for these faces.

INSTALLATION OF THE GATE

31. Gate items which required accurate positioning within the caisson structure included stainless steel landing plates, the main trunnion brackets within which the gate pivots rotate, upper and lower locking bolts each side of the entrance and push-off rams. It was also necessary to position the 'quayside' sheaves which carried the main operating rope to align with the gate sheaves after installation of the gate. All of the above equipment was installed, together with the two sets of gate machinery, within an intensive period of activity during June and July 1991. To achieve this required continued co-operation between the civil contractor, Monk/Ballast Nedam, and the gate contractor Sir William Arrol, and regular liaison and supervision by ABP.

32. To facilitate installation of the dock gate, a removable panel was incorporated in the internal temporary bulkhead. This would allow flooding of the caisson and floating-in and installation of the gate whilst retaining the option of subsequently pumping the caisson dry should problems arise. The gate was installed and tested at the end of August 1991. Stepping the gate was carried out by ballasting the lower buoyancy tanks and using cranes and winches to position the gate trunnion pivots on to the trunnion brackets assisted by a diving team. Following installation, a seal test was implemented by pumping out between the gate and the outer temporary bulkhead. Fig.13 shows the gate installed and excavation for the outer apron proceeding with the outer bulkhead still in place.

COMPLETION

33. Construction of the aprons, removal of the outer bulkhead, completion of dredging and commissioning of the gate machinery were sufficiently advanced to allow the opening of the new entrance to shipping on 15 October 1991. A period of 'dual operation' followed, when both entrances were available for use, until 30 October when the contractor was allowed to commence work on the closure of the existing entrance.

34. The inner dock wall of the closure was constructed as a concrete block gravity structure and rock armouring slope protection was used on the outer face, the main fill material being sand excavated from the site of the caisson. The lead-in jetty comprised steel tubular piles, concrete deck and energy absorbing fenders. The deck was largely precast in order to expedite construction and overall completion of the project was achieved on 5 March 1992 (Fig.14).

NEW DOCK ENTRANCE

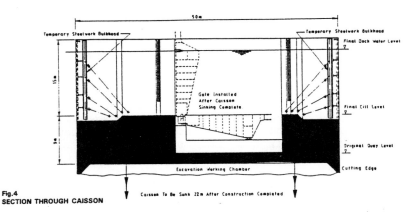

Fig.4
SECTION THROUGH CAISSON

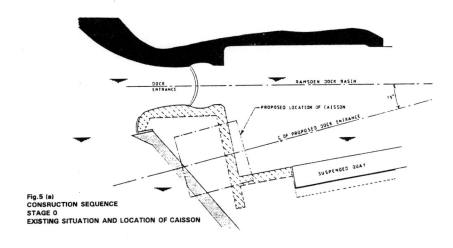

Fig.5 (a)
CONSRUCTION SEQUENCE
STAGE 0
EXISTING SITUATION AND LOCATION OF CAISSON

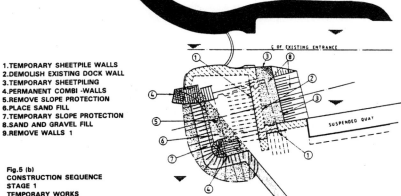

1. TEMPORARY SHEETPILE WALLS
2. DEMOLISH EXISTING DOCK WALL
3. TEMPORARY SHEETPILING
4. PERMANENT COMBI-WALLS
5. REMOVE SLOPE PROTECTION
6. PLACE SAND FILL
7. TEMPORARY SLOPE PROTECTION
8. SAND AND GRAVEL FILL
9. REMOVE WALLS 1

Fig.5 (b)
CONSTRUCTION SEQUENCE
STAGE 1
TEMPORARY WORKS

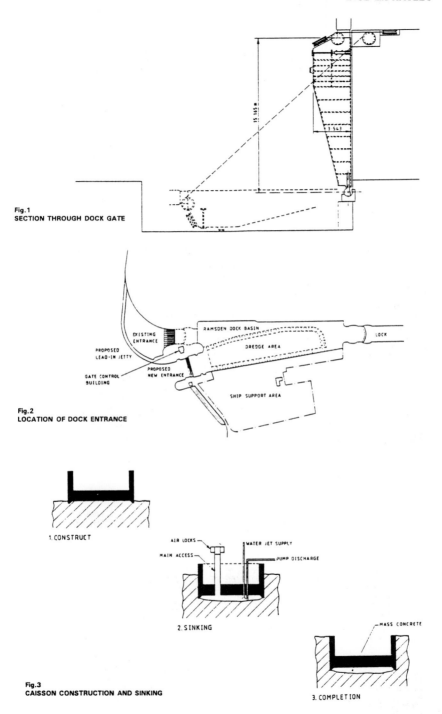

Fig.1 SECTION THROUGH DOCK GATE

Fig.2 LOCATION OF DOCK ENTRANCE

Fig.3 CAISSON CONSTRUCTION AND SINKING

NEW DOCK ENTRANCE

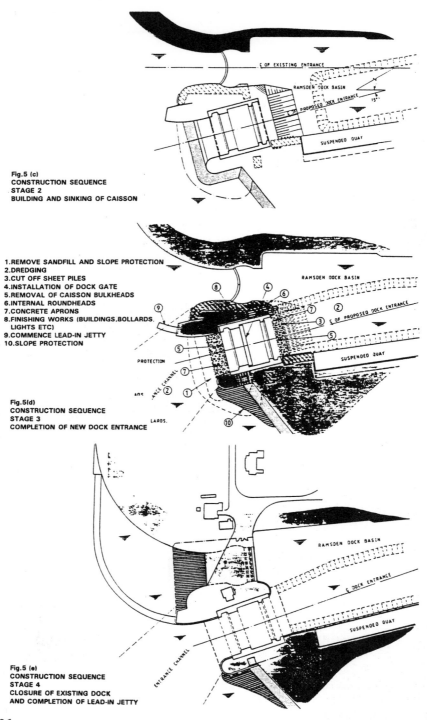

Fig.5 (c)
CONSTRUCTION SEQUENCE
STAGE 2
BUILDING AND SINKING OF CAISSON

1. REMOVE SANDFILL AND SLOPE PROTECTION
2. DREDGING
3. CUT OFF SHEET PILES
4. INSTALLATION OF DOCK GATE
5. REMOVAL OF CAISSON BULKHEADS
6. INTERNAL ROUNDHEADS
7. CONCRETE APRONS
8. FINISHING WORKS (BUILDINGS, BOLLARDS, LIGHTS ETC)
9. COMMENCE LEAD-IN JETTY
10. SLOPE PROTECTION

Fig.5(d)
CONSTRUCTION SEQUENCE
STAGE 3
COMPLETION OF NEW DOCK ENTRANCE

Fig.5 (e)
CONSTRUCTION SEQUENCE
STAGE 4
CLOSURE OF EXISTING DOCK
AND COMPLETION OF LEAD-IN JETTY

Fig. 6

Fig. 7

Fig. 8

NEW DOCK ENTRANCE

Fig. 9

Fig. 10

Fig. 11

Fig. 12

Fig. 13

Fig. 14

Disposal of dredged spoil and channel design

Captain J. W. GREEN, Port Manager & Harbour Master, Barrow and Silloth, Associated British Ports

SYNOPSIS. In order for the new 'Vanguard' Class submarines being built at Barrow-in-Furness to safely enter and leave the port it was necessary to deepen the approach channels. This required a major dredging operation together with a suitable disposal site for the dredged material. It was also necessary to confirm that the channel was adequate for this new class of submarine, which was significantly larger than any that had been built previously in the United Kingdom. This paper describes some of the main actions that were required to achieve a suitable disposal site and a properly designed channel.

DISPOSAL OF DREDGED SPOIL
DISPOSAL OPTIONS
Introduction
1. To ascertain whether the 8 miles approach channel to the Port of Barrow could be dredged to the required depth for Vanguard Class submarines Wimpol Ltd. carried out a geophysical and vibro-coring survey in 1987. The estimated volume of material to be dredged was over 3,500,000 cubic metres and the existing disposal sites were inadequate, hence a new disposal site was necessary.
Limitations of Existing Disposal Sites
2. Barrow had two disposal sites, Site 'B' a circle of $\frac{1}{4}$ mile radius for capital spoil in 11 metre water depth and site 'A' a semi-circle of $\frac{1}{2}$ mile radius in $3\frac{1}{2}$-7 metre water depth. The material from the former tended to return to the channel and the latter had limited capacity and both were therefore unsuitable, although they had the advantage of being within close proximity to the channel
Alternative On-shore Disposal Sites
3. The dredge tenderers were invited to investigate alternatives to disposal at sea and various options were actively investigated. Numerous environmental and financial obstacles were encountered which made the schemes uneconomic.
Requirements for the Sea Disposal Site
4. There were 4 main requirements for the sea disposal sites:

i) The dumped material should not return to the dredged channel
ii) The sites should have adequate capacity for both the capital and ongoing maintenance dredge requirements.
iii) The sites should be near to Barrow.
iv) Two sites should be available to avoid conflict between different dredging contractors.

DISPOSAL LICENCE
Liaison with MAFF
5. Because the Oslo Dumping Convention places restrictions on new disposal sites MAFF were given as much notice as possible of the disposal requirements as were the Department of Transport Marine Division and the Crown Estate Commissioners.
6. It was suggested to MAFF that site 'A' could be extended beyond the 10 metre depth contour which would give adequate capacity without causing shoaling hazards to shipping. The MOD Hydrographer was consulted and confirmed the sites suitability if a minimum water depth was maintained. Other sites were also considered.(Fig.1)
7. MAFF liaised with the MOD Hydrographer, the Department of Transport, the Nature Conservancy Council, the fishing industry and other local interested parties, in order to ensure the proposed site was suitable.
Offshore Site Investigations
8. MAFF'S four main concerns regarding the offshore site were:-
i) The material to be dumped should not be contaminated.
ii) The movement of material after dumping.
iii) The effects on benthic animals and fishing.
iv) Safety of navigation.
9. Samples for analysis as to contamination and dredgeability were taken from the surface and a depth of 2 metres at 20 equidistant locations in the channel between Ramsden Dock Entrance and Lightning Knoll by a small jack-up barge. Samples from 10 locations were sent to MAFF and the results of the core analysis given to the prospective dredging contractors.
10. The MOD Hydrographer provided MAFF with tidal stream, current and wave data plus historical and contemporary soundings of the area. MAFF undertook a site survey including current recording and release of sea-bed drifters. Delft Hydraulics were also commissioned to advise on the suitability of the site and they agreed with MAFF that the net sediment transport would be seaward and the amount of disposed material in suspension was a very small percentage of the total load circulating in Morecambe Bay.
11. MAFF required the dump site to be as small as possible to minimise the blanketing effect and in December 1990, two years after the initial request was submitted a revised disposal site plan was sent to MAFF. The site was seaward of the 10 metre

depth contour, clear of the main shipping routes with depths of 7.1 metres to the south east so that a minimum depth requirement of 7.0 metres was acceptable. In practice shoaling has been experienced with depth reductions to 8 metres but this has not been a problem.

Licence conditions

12. The disposal licence was issued in January 1991 about 3 months before commencement of the capital dredge and in addition to the usual licence conditions. there were some special requirements which were:
- i) The spoil was to be spread evenly and a minimum depth of 7.0 metres maintained.
- ii) Quarterly surveys of the site were required.

MONITORING OF DISPOSAL OPERATIONS

Control of Dumping Operations

13. In order to ensure an even spread of material in the dump site the disposal vessels were fitted with a track plot device recording the vessels position between the start and end of each dumping operation. Short range radio position fixing was used with signals transmitted from fixed beacons which were also used by the dredgers and survey craft. Any irregularities were investigated and the system was very effective with the trailer dredgers, although prone to operator error with the smaller hopper barges. A dredging inspector was employed to carry out general overseeing of the dredging and disposal operations.

Survey of Dump Site

14. A pre-dredge survey of the dump site was carried out which confirmed no shoaling had taken place since the original survey to assess its suitability. Monthly surveys were undertaken during the period of the capital dredge to monitor shoaling and two shoal areas did develop which were controlled by enforcing an exclusion zone round them. The dredging was carried out by a Joint Venture comprising 3 Companies and they subdivided the dump site between themselves to ensure effective accountability for even spreading of the material. Post dredge surveys were undertaken and these confirmed the depths in the dump site were satisfactory.

CHANNEL DESIGN

Introduction

Submarine Construction at Barrow

15. VSEL built the Royal Navy's first submarine 'Holland I' at Barrow in 1901 and they are now building the latest 'Vanguard' Class which will carry Trident missiles, as Britain's nuclear deterrent. With 'Vanguard's' increased size and draught it was decided a wider entrance was needed for the port together with increased depth of water out to the open sea. The channel Design work was required to ensure that 'Vanguard' could safely navigate in and out of the Port of Barrow.

Options for Moving Vanguard class Submarines between the Port of Barrow and the Sea

16. The options for moving 'Vanguard' were either to use a transporter vessel or for the submarine to remain afloat. Eight alternatives were considered before deciding on self propelled transits, having verified the channel was physically suitable and all aspects of nuclear safety were satisfactory.

Vanguard Class Requirement

17. To achieve a self propelled exit VSEL and the MOD chose to build a deeper and wider dock entrance and deepen the approach channel out to Lightning Knoll. The entrance would have a single flap gate, which would lower on to the dock bottom and provide a better seal than traditional gates and not be vulnerable to damage when open.

Statutory Requirements

18. In 1987 R. T. James & Partners a Civil Engineering Consultant drew up the conceptual design of the new dock entrance for VSEL and the MOD and produced plans to show the limits of deviation which were required for a private act of Parliament to authorise such works in tidal waters. The Bill was the 'Associated British Ports (Barrow) Act 1988 which gave ABP powers to control navigation and dredge the Walney Channel and approaches out to Lightning Knoll, to compulsorily purchase land for the works, and instal navigation aids subject to the sanction of Trinity House and approval of the Department of Transport under the Coast Protection Act 1949, Section 34. The disposal of dredgings also required the Department of Transport's consent and that of the Crown Estate Commissioners under the Coast Protection Act 1949 together with approval from MAFF under the Food and Environment Protection Act 1985, Part II.

19. Work on installation and demolition of the navigation aids in the S.S.S.I. surrounding the channel involved consultation with the Nature Conservancy Council as required by the Wildlife and Countryside Act 1981. Early consultation was essential as it enabled work in the nesting sites to be done outside the breeding season.

20. Barrow B.C., N.R.A., N.W. Water and British Gas were all advised of the works and the latter were fully involved in the contractual arrangements and dredging operations in the vicinity of their 36" diameter gas pipeline under the Walney Channel.

21. Notices to Mariners were promulgated to keep seafarers aware of the works and the contractors made their own arrangements with landowners for wayleaves and access.

NEW DOCK ENTRANCE AND DEEP DREDGED CHANNEL

Improved Operational Window

22. The new dock gate can be opened about $2\frac{1}{2}$ hours before high water when the tide height is just over $5\frac{1}{2}$ metres and can be kept open if necessary up to $1\frac{1}{2}$ hours after high water on most tides. The sill level is 3 metres below the lowest

astronomical tide (L.A.T.), the latter being the sill level of the old entrance, and the channel out to Lightning Knoll was deepened by a similar amount which gave a greater operational capability.

23. Shipping is vulnerable to the powerful forces of the sea and the marine environment is an ever changing one due to the influences of barometric pressure, wind, waves, tides and currents, which effect both the sub sea and surface environments. These influences are constantly changing, which makes ship handling very much an art as well as a science and in confined tidal waters the Pilot is all the time assessing how the combined effects of wind, tides and currents will effect ship handling so that he can adjust the course and speed accordingly.

24. It is essential to maintain adequate underkeel clearance to (i) avoid bottom contact in waves causing vessel movement and (ii) shallow water effects which can impair steering. Before making a transit the pilot will always assess the weather and tidal conditions together with the water depth to ensure they are suitable for a safe passage. The purpose of the Channel Design was to ensure that all the physical requirements were in place and the variable parameters identified and measurable so that transits of the channel could be carried out safely by 'Vanguard' Class submarines.

25. Submarines handle very differently to surface ships due to their different underwater shape and rudder/propulsor configuration. Submarines are directionally very stable and have little or no transverse thrust from the propulsor. The windage on the superstructure is also different with the fin acting like an aerodynamic vertical sail, and the casing being low and rounded offers little wind resistance.

EXISTING DATA
Boat Characteristics
26. The draught and trim of 'Vanguard', together with it's handling characteristics from free running and captive model tests in waves and still water had been determined by A.R.E. (Haslar) in test tanks. These tests also provided data on acceleration, deceleration, turning, squat, heave, shallow water and bank effects.

27. The Port of Barrow to the north of Morecambe Bay is sheltered by Walney island and was first established in the 12th century. Major dredges were carried out for the battle cruiser 'Princess Royal' in 1911 and the first nuclear submarine HMS 'Dreadnought' in 1962/63, with ongoing maintenance dredging in between times. Dredging and survey records were given to the channel designer together with a mass of data on winds, tides, surges, currents, waves, charts, soil conditions, geology, ecology, research reports, navigation aids. There was little wave data and where there was a data shortfall more investigations were necessary.

DATA REQUIRED
Adequacy of Proposed New Channel
28. In order to assess the suitability of the proposed channel all the environmental influences and physical characteristics of the channel and the submarine needed to be combined in one comprehensive model. The existing channel widths of 90 metres, 100 metres and 140 metres had been retained in the Walney, Inner and Outer Channels respectively with a departure time of 35 minutes before high water on sailing to coincide with a lull in the tidal stream at the dock entrance and an arrival time of 20 minutes before high water on entry to arrive on a rising tide with some margin of safety.

29. Tide gauges were being installed at Roa Island and Halfway Shoals to provide data and real time information on relative tide heights along the channel. Field studies were also necessary to give accurate information on siltation, and the channel was re-aligned to take advantage of the natural deeps to minimise both capital and maintenance dredging costs.

Influence of the New Channel on the Submarine
30. Deepening the channel would significantly increase both its' cross sectional area and the water volume being moved by the tidal currents. It was essential to determine what effects these changes would have on the handling of the submarine and the relative tide heights and currents in the channel.

Navigational Requirements
31. The 8 mile approach channel to Barrow has depths in excess of 5 metres below L.A.T. at Lightning Knoll which gradually decrease to 3 metres below L.A.T. at Ramsden Dock Entrance. The tide heights above L.A.T. vary between 6.3 metres and 10.3 metres on the lowest and highest tides respectively, these heights being subject to variation with tidal surges. The channels are adjacent to shoal areas which dry to varying degrees beside the Inner and Walney Channels.

32. The navigation aids were designed to keep vessels within the toe lines of the channel under all circumstances with an element of redundancy. The main guidance is provided by transit beacons aligned with the centre lines of the channels, with additional 'gated' beacons midway along the Walney Channel and at the south end of the Inner Channel. To assist vessels approaching or leaving the Ramsden Dock Entrance two additional beacons were installed on the north east side of the Walney channel in this area, and transit beacons plus an Inogon light were installed to guide vessels through the entrance itself. Gated buoys with lights and radar reflectors were installed along the length of the channel and these together with the channel beacons are located 50 metres back from the toe-lines to allow for dredging operations and minimise the risk of collision with passing ships. P.E.L.'S (Port Entry Lights) with coloured sectors were installed on No.5 transit beacon to guide vessels along the slight deviation in the Walney Channel over a gas pipeline, and near to the dock entrance where No.10 beacon does not have a rear transit. (Fig.2)

33. Because of the long approach from Lighting Knoll an additional beacon with a Racon was installed in the vicinity of Halfway Shoal to give additional guidance in this area and to assist with vessels making a landfall at Barrow.

34. The Barrow Pilot together with the Commanding Officers and Navigators of submarines under construction in the port were involved in the design of the navigation system and their recommendations included.

CHANNEL DESIGN
Introduction

35. The channel design necessitated the combination of all the varied and complex environmental and physical data together with the characteristics of the submarine into one comprehensive study. In view of the important strategic and safety aspects of the project only organisations of international standing with a proven record of achievement were invited to quote for the work, and Delft Hydraulics (DH) were the successful tenderer.

Delft Hydraulics Tasks
Task 1 : Evaluation of Existing Data

36. Adequate existing data was included in the respective task and any shortfall was made good.

37. DH reviewed the wind data, offshore wave climate, tides, currents and negative surges, sedimentation/erosion patterns and bottom material, the submarine hydrodynamics and soil conditions which were considered adequate except for swell data, complete wave spectra data, effects of bottom friction and wave breaking along the channel, and certain aspects of the submarine hydrodynamics. Additional field studies were undertaken to verify the currents in the channel, sedimentation rates from a trial dredge and some additional tank test on the submarine model were carried out to assess its movement in waves.

Task 2 : Site Investigation

38. DH took additional tidal readings at Lightning Knoll and Haws Point East and current measurements at Lightning Knoll, Roa Island and Ramsden Dock Entrance area together with drogue tracking in Piel Channel and in front of Ramsden Dock. Vertical current velocity profiles and suspended sediment concentrations were measured at 5 different depths at Halfway Shoal, off Roa Island, the west end of the Walney Channel and off Ramsden Dock Entrance, the latter being analysed in an on-site laboratory. Cross section soundings were taken at the same locations. Salinity and density measurements were taken at numerous locations with little variation and 67 bottom samples were taken with a 2 litre Van Veen grab, which were analysed for grain size distribution. Bedform tracking was carried out along the axis of the channel using a Ratheon DE-719 echo sounder and visual observations were made of wind and waves.

Task 3 : Tides and Currents

39. A consistent data set of currents and water levels was required for the seaward boundary of the Walney Channel Model to be used for the siltation studies and the submarine transit simulator. This was achieved using a suite of 3 nested flow models of the Irish Sea, Morecambe Bay and Walney channel. Admiralty charts provided bathymetry and Ordnance Survey maps coastline definition, tidal and current data being obtained from numerous recording stations. The model grid sizes were progressively smaller and decreased towards the coast to give greater detail.

Task 4 : Wave climate

40. Wave climate data was required for siltation predictions and for the channel width and depth requirements. No long term wave data was available but limited data was available for Morecambe Bay Light Vessel, Lightning Knoll, Halfway Shoal and the BMO hindcast point which was analysed to derive wave height exceedance, directional distribution and the relationship between height and period. At Halfway shoal the dissipation of wave height by bottom friction and wave breaking was evident, as well as refraction.

41. Wind data from various locations in the area was statistically analysed to produce wind roses, and wave hindcast computations were made for directions with short fetch lengths. The channel wave climate was deduced, using wave propagation computations from the offshore wave climate after allowance for local factors because of the lack of correlation between significant wave height and period. Winds from the south produced lower waves than winds from the south west because of the formers shorter fetch, and winds from south round to west with a longer fetch and period were more susceptible to refraction, wave breaking and bottom friction than those from north through east to south.

42. DH identified swell waves emanating from storms in the south west Irish Sea or beyond as being significant because of their long period and effect upon vessel movement.

Task 5 : Detailed Currents and Sedimentation

43. The capital dredge of the Barrow channels will have a significant effect on the hydraulic and morphological regime and the geometry of the channel will react accordingly, the rate of change being dependent upon the sediment transport rates and the transport gradients. DH used mathematical models to analyse these effects with input data on currents, waves, sediment characteristics, transport rates and deposition volumes. This information was required to determine the amount of maintenance dredging and an efficient dredging strategy, together with adequate siltation capacity in the design depth of the channel.

44. Each section of the channel had distinct and different characteristics and DH used 4 computer models to simulate water levels, currents, sediment transport processes and the irregular bathymetry together with the effect of suspended

sediment on non-erodible bed layers. DH also investigated the alternative disposal sites for the dredgings.

45. DH predicted there would be little change in tide heights but currents would increase off Haws Point and decrease in the Walney channel.

46. DH calibrated their models from the real time development of their trial dredge pit and from the historic dredged volumes, as well as data collected from their field studies. Computations were made with a number of combinations of wave and current conditions, using a weighting to take account of the occurrence of each condition to produce annual average sedimentation rates with seasonal variations. DH cautioned however that the annual volume results could be a factor of 1.5 too low to a factor of 2 too high, and the outer exposed channel could be vulnerable to rapid infill in storms.

47. DH confirmed that the capital dredge would not have an adverse effect on Walney Island or the surrounding areas and they recommend the disposal site south of Lightning Knoll as a suitable location.

Task 6 : Channel Depth

48. This task determined the nautical depth for the channel allowing for the vessel's draught, squat and motions together with tide and waves with an acceptable safety level and downtime. Adequate siltation buffers were necessary allowing for survey inaccuracy together with an appropriate maintenance dredging strategy. The costs of capital and maintenance dredging, channel downtime costs and safety levels also needed to be considered and altogether this required deterministic and probabilistic methods to integrate the various elements.

The inputs required were the channel layout, tide curve, wave climate, squat, ship motion characteristics and safety criteria for each section of the channel. (Fig.3)

49. The computations involved the use of 3 models to determine the underkeel clearance (U.K.C.), the probability of bottom contact and the tidal windows respectively. The vertical ship motion calculations used the results from tank tests at A.R.E.(Haslar) and HR Wallingford with different wave directions, heights and periods input to the DH computer model although it was not possible to reproduce the damping effect of the after planes. Ship motions are most affected by the period of the waves and inward of Piel Harbour waves were not significant. The effect of long period swells from St. Georges Channel for instance could be very significant and DH recommended closure of the channel with swell waves present over 0.15 metres. The effects of free and bound long waves also needed to be considered.

50. Tidal surges have effect over the length of the channel with tide heights being effected by barometric pressure and wind direction and force.

51. Each ship type has a safe minimum and average U.K.C. and DH calculated these for the submarine so that it could manoeuvre safely. Practical precautions such as entering on

a rising tide were also incorporated into the design, together with an assumption on the acceptable period of channel downtime, from which DH arrived an optimum channel depth with limiting wave and tidal parameters.

Task 7 : Channel Width

52 To determine the channel widths DH used their real time ship simulator with a computer generated outside view. All the ship and environmental data was input to the simulator and tests were performed for a wide range of conditions the variables being tide, currents, waves, wind, transit direction, submarine alignment and area of the channel. The bridge controls were provided such as radar, gyro compass, log, rudder indicator, wind speed and direction indicator, RPM indicator etc. and the outlook reproduced over an arc of 70° looking forward, together with a plot of the ships position and alignment in the channel.

53. Rudder and engine movements were input to the computer which calculated their effects on the submarine's speed, track and heading taking account of current, wind, waves, water depth, bank and shallow water effects. Plots of the runs were produced and statistically analysed. The indications were that the effects of wind and wave height were not significant and there was little difference with mean and spring tides, but a rate of turn indicator would be beneficial on the bridge.

54. The proposed channel was confirmed as being suitable as were the intended Voith Schneider tugs.

Task 8 : Dredging Studies (Undertaken by DHV)

55. The objective of this task was to recommend an efficient dredging strategy consistent with a stable channel and side slopes and adequate siltation buffers. The channel area is a complex site of sands, gravels, boulder clays and cobbles in open sea to sheltered water conditions with a wide range of wind, sea, current and tidal conditions and these variables made accurate results difficult.

56. DHV considered the probabilities of the significant winds and waves together with the range of tides and currents and the various soil types as well as survey inaccuracy. Side slopes of 1:10 were recommended particularly because of wave action in the outer reaches and this in effect was achieved by dredging outside the toe-lines to give a side buffer which allowed for slope degradation.

57. A maintenance dredging strategy is a combination of plant, frequency and organisation and at Barrow with its sea disposal site a trailer suction dredger is the only cost effective option. The siltation buffers allowed is a balance between capital and maintenance dredging costs, siltation rates, minimum depth requirements, whilst leaving a practical working depth tolerance for a trailer which is normally a minimum of 0.5 metres. Horizontal siltation buffers are also advisable where bed load and gravity infill is expected.

58. DH calculated 80% of the siltation would be in the Outer Channel caused by cross currents, littoral drift and wave

action, and in extreme storm conditions the infill at Halfway Shoal could be 20% of the annual total. There would also be season variations in the upper Walney Channel with more siltation in the winter due to higher winds and waves.

59. Because of the intermittent requirement for the maintained channel a larger more economical dredger could be used but draught must be compatible with the water depth to minimise tidal downtime, and an adequate period must be allowed to achieve the depth requirements on each occasion. Relative costs were provided for different dredging strategies with allowance for anticipated volumes of dredged spoil, efficiency and weather downtime etc.

Task 9 : Channel Design

60. DH's task was to optimise the channel dimensions, the depth and width of which would allow the safe passage of Vanguard Class submarines, taking account of all previous studies and investigations. It was recommended a tidal window of 3 consecutive days should be allowed to minimise the risk of delays of weeks rather than days.

61. Due to the nature of the adjoining areas DH confirmed there was little scope for altering the alignment of the channel and the variables were the siltation buffers and the wave and tidal constraints that could be imposed. The nautical depth was determined from two sets of parameters the fixed ones related to the submarine such as squat and ship motions and the variable ones such as waves, tides and surges.

62. MOD had originally set a tidal threshold for transits and subsequently accepted certain wave height constraints in order to limit costs. A negative surge allowance of 0.3m was added to the depths and in order to assess the various alternatives DH calculated the channel downtime at various tide heights and wave heights and the channel depth implications for each scenario. There was a difficulty in so far as there was little data on the persistence of waves.

63. The channel depth comprised components for nautical depth required by the submarine, survey inaccuracy, siltation buffers. including ripples and storm infill, and overdredge to provide a maintenance dredge zone. DH calculated the siltation buffers for 2 months in each section of the channel to which the other components of the depth were added from their investigations to arrive at the final channel design depth. A minimum siltation buffer of 0.6m was recommended with depths specified for a minimum channel length of 1 kilometre. The dredged depths were based on an annual siltation level of $2,450,000 m^3$.

64. Because of the intermittent requirement for the dredged channel over the life of their submarine build programme the MOD chose the reduced cost option which allowed siltation above the design level.

Task 10 : Navigation Aids

65. DH recommended the radio position fixing system with a display showing the submarines attitude in relation to the

transit line together with a rate of turn indicator should be the primary navigation system. The transit beacons would be the secondary system with the buoys and beacons being the tertiary system, except in the Outer Channel during an exit where the roles of the secondary and tertiary system would be vice versa.

Task 11 : Impedance Procedures

66. DH recommended procedures to be adopted in the unlikely event of a main propulsion failure during a transit requiring the completion of the operation under tow. Towing speeds were identified and trim conditions for towing ahead or astern.

Task 14 : Monitoring

67. DH took current readings at different depths and locations off Ramsden Dock Entrance after the capital dredge to compare the actual current fields with their model predictions and these showed good comparison. Currents in this area are most significant because of the need for the submarine to manoeuvre in and out of the entrance across the tidal stream which continues to run north for more than an hour after high water.

Task 15 : Achievement of Design

68. The purpose of this task was to verify the dredged channel conformed to the design, the navigation aids were in the correct locations and fulfilled their purpose and the actual sedimentation rates compared with the predictions.

69. A DH representative was in attendance during the post dredge survey to independently verify the channel depths and widths and also to audit the survey methods to ensure they achieved the required degree of accuracy. Where the channel layout did not conform for practical reasons to the DH design this was acceptable to DH with certain proviso's which were agreed.

70. The navigation aids were satisfactory, although for practical reasons it was not possible to place the buoys on the channel toe-lines as DH requested, because they would have impeded dredging operations and they were placed 50m back from the channel boundaries.

71. Sedimentation was comparable with DH's predictions except in the outer reach where in the non-erodible layer areas where the transport capacity exceeds the erodible sediment available and siltation was less than anticipated.

72. Prior to submarine movements the navigators and tug masters go through a training course arranged by DH on a ship simulator, which the participants found very useful.

73. The tidal regime at Barrow is continuously monitored by the 3 tide gauges and the data analysed by the MOD Tidal Branch.

74. Nuclear Safety procedures cover all aspects of submarine movements and VSEL continually monitor operations to ensure procedures are followed.

75. Wave data is provided on days of transits and a Meteorological Office weather forecaster is on site before and during transits to ensure expert advice and real time

information is available on all environmental aspects.

76. Maintenance dredging campaigns have been carried out prior to each submarine transit and these have been successfully completed despite some extended periods of bad weather at times.

77. HMS 'Vanguard' completed her sea-trials in 1993 and has been commissioned into the Royal Navy thus successfully concluding the channel design project.

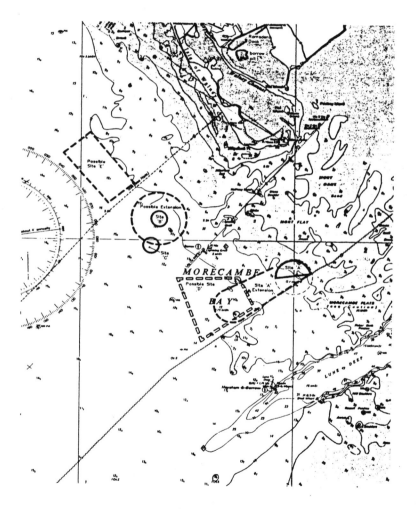

Fig.1 Proposed Spoil Disposal Sites

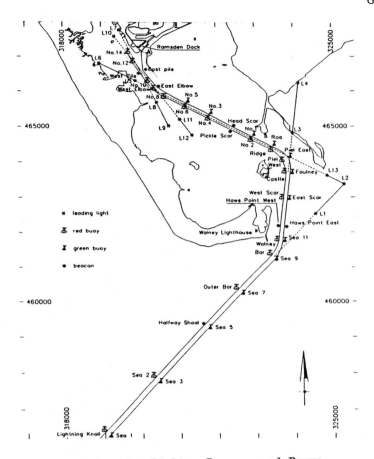

Fig.2 Leading Lights, Beacons and Buoys

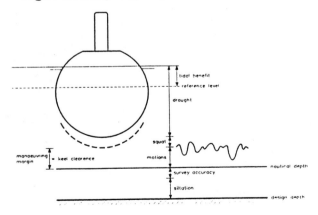

Fig.3 Components of Channel Depth

Dredging for Trident class submarines

R. J. GARDNER, Business Development Manager, and D. H. ROBERTS, Director (retired), Westminster Dredging Company Limited

---

SYNOPSIS. The dimensions of Trident Class submarines under construction at Barrow-in-Furness in Cumbria required major improvements of Walney Channel to provide for their safe navigation. This requirement resulted in Associated British Ports (in co-operation with Vickers Shipbuilding and Engineering and Ministry of Defence) issuing invitations to tender for "Dredging and Deepening the Channel from Ramsden Dock Entrance to Lightning Knoll and Other Associated Works". Tender invitations were issued on 30 April 1990 to a selected list of international dredging contractors, all of whom had a significant level of experience in the UK dredging environment.

THE JOINT VENTURE
1. At an early stage in the pre-tender period, the dredging companies D. Blankevoort, Jan de Nul (UK) and Westminster Dredging Company recognised the advantages of forming a joint venture for the proposed execution of this major dredging contract, and a legally binding joint venture was constituted with Westminster Dredging, by virtue of its long term experience of working in Walney Channel, fulfilling the role of co-ordinating partner.
2. The contract as tendered for involved a limited execution programme time, an imperative completion date with high level of liquidated damages, the employment of multiple plant resources, and the provision of adequate back-up resources.
3. The available resources of the partners included heavy mechanical dredgers, a fleet of medium range shallow draughted trailer suction dredgers, the world's largest cutter suction dredger and an adequate range of back-up and support plant. The individual status of the partner companies was such that the risks inherent in the timely delivery of the works were virtually eliminated.
4. The joint venture tender was submitted to ABP on 15 October 1990 and subsequent to the submission of two further tenders to embrace modifications and clarifications, the works were awarded to the joint venture on 28 March 1991.

THE TENDER DOCUMENTS

5. The Tender Documents called for fixed lump sum price to cover execution of the Works with substantial completion on 31 December 1991. However, in recognition of the protracted pre-award negotiations, completion date was deferred to March 1992.

6. Tender conditions included identified clauses of G.C.Stores 1, Defcons, ICE 5th (rev.'79), Special Conditions, and Quality Assurance and Control requirements deriving from AQAP(NATO) and Defstans. The special conditions identified ABP as the employer and required the successful contractor to design, construct, complete and maintain the works. However, the maintenance obligation did not apply to dredging work.

7. Although substantial soils information was provided, it was the clear responsibility of the contractor to dredge in any material.

8. In general, the tender documents gave clear definition of the scope of works and gave clear identification of the responsibilities and obligations of the successful tenderer.

BRIEF DESCRIPTION OF THE WORKS
Dredging

9. For ease of reference, the section of channel between Ramsden Dock entrance and Bar Buoy was referred to as the inner channel, and the section between Lightning Knoll and Bar Buoy was referred to as the outer channel. Inner channel length was approximately 8km and outer channel length was approximately 7km, giving a total length of 15km. Design width between toelines was variable for practical navigation reasons on a general basic width of 120 metres. Existing advertised depths were in the following depth ranges:-

     (a) Ramsden Dock entrance to Pickle Scar   -0.80m LAT
     (b) Pickle Scar to Bar Buoy   -1.70m LAT
     (c) Bar Buoy to Lightning Knoll   -2.30m LAT

However, some areas of the inner channel and substantial areas of the outer channel were already significantly deeper than the above levels.

10. The contractual design depths involved a generally progressive deepening of the channel seawards from Ramsden Dock Entrance, although variations in this progression were required in limited areas to provide siltation buffers. Required minimum design depths were as follows:-

Inner Channel

Ramsden Dock Entrance	-3.30m LAT
Buoy No. 11 (1,000m seaward of RDE)	-3.20m LAT
Buoy No. 7	-3.17m LAT
Area of gas pipe crossing	-3.54m LAT
Pickle Scar	-3.42m LAT
Coup Scar	-3.60m LAT
Channel Scar	-4.35m LAT

Outer Channel

Bar Buoy	-4.52m LAT
Halfway Shoals	-5.87m LAT
Lightning Knoll (inshore)	-5.50m LAT
Lightning Knoll Buoy	-5.33m LAT

The main gradient of side slopes was 1:8 with limited sections of the slopes of the inner channel being restricted to 1:5 to preserve the stability of the training walls. Whereas pre-dredge survey information was provided by ABP, determination of volume to be dredged was the contractor's responsibility.

11. ABP made a MAFF approved sea deposit site available for the sea disposal of dredgings, situated 2 nautical miles to the south of Lightning Knoll Buoy and entailing a round trip sailing distance of 18 nautical miles from Ramsden Dock Entrance. The area of the site was approximately 4,000,000 square metres and the dump capacity was in excess of 10,000,000 cubic metres. Limitations applied to the deposit of dredged material and graphic print-out of every dump position was a MAFF requirement.

MATERIAL TO BE DREDGED

12. Substantial soils information was made available by ABP, further substantiated by 20 boreholes taken in the channel during the tender period. Limited trial pits were also dredged and the joint venture executed some independent soils investigatory works, primarily to verify existing information.

13. In principle, the material comprised sand, silt, gravel, cobbles, boulders, and boulder clay. Concentrations of boulders were anticipated, and occurred, in the outer channel and the possible incidence of very large nomadic boulders was recognised. Boulder clay was a common feature of the outer channel works and it occurred significantly at deeper levels in the inner channel. The granular material ranged between loose and very dense and clays up to the "hard" classification were apparent. There was no evidence of rock in any of the soils information.

14. ICE Conditions 11 and 12 were either modified or eliminated from the applicable Conditions of Contract, thereby allocating the soils risk to the contractor. Although the value of the available soils information was significant, it became clear that dredging evaluation of the soils would involve speculation in the various areas of channel which were not covered by the investigatory work. Soils evaluation thus constituted one of several serious risk areas to be considered by the joint venture.

THE TENDER PREPARATION

15. During the tender period, the economy of shore disposal of dredgings was closely investigated by the joint venture as an alternative to sea disposal. Whereas financial and other benefits were apparent, the greater part of the environs of the inner channel lies within Sites of Special Scientific Interest and the failure to achieve harmonious compatibility with Nature Conservancy Council and National Rivers Authority (in their protection of the littoral environment) led to the abandonment of this proposed method of disposal of dredgings.

16. Sea disposal techniques therefore influenced the joint venture's choice of dredging plant, and it was decided to

employ trailer dredgers for the dredging of the loose sedimentary deposits both in the inner and outer channels, heavy mechanical dredgers for the cohesive soils in the inner channel, and very large cutter suction dredger with barge loading equipment for the heavy excavation work of the outer channel.

17. As the Works were not subject to remeasure, the accuracy of dredged quantity calculations inclusive of vertical and horizontal tolerances and the influences of interim siltation was of prime importance.

18. The Institute of Sound and Vibration Research of Southampton University was commissioned to assess noise emissions from the mechanical dredgers and following this, discussion and agreement with the Environmental Noise Officer of Barrow Council was reached in respect of hours of work proposed for the mechanical dredgers (in the inner channel) relative to noise emissions.

19. Wayleaves for haul roads and electric cable routes to the beacon structures in the inter-tidal zone of the inner channel were agreed with all relevant land-owners and with Nature Conservancy Council and related bodies in areas classified as Sites of Special Scientific Interest.

20. Weather statistics were evaluated from long term records of Westminster Dredging's previous works at Walney Channel and applied to the working characteristics of the proposed plant, being particularly influential on the inner channel plant using the sea deposit site, and on the works of the large cutter suction dredger working the outer channel.

21. Risk evaluations involved consideration of:-

(a) Soils and weather conditions (as stated above).
(b) Average overdredge ranging between 0.30m deeper than design levels for mechanical dredgers and 1.00m deeper than design levels for the large cutter suction dredger working in the open sea conditions of the outer channel.
(c) Side slope stability with the contractor taking responsibility for the stability of side slopes and any additional dredging resultant from instability. The joint venture accepted unlimited responsibility for side slopes over a total length of 25,000 metres, but declined responsibility for any influences of instability on the upper inner channel training walls over a total length of approximately 5,000 metres.
(d) Interim siltation during the course of dredging operations. This was more readily predictable for the inner channel but was essentially speculative for the outer channel where the littoral drift down the coast of Walney Island is in a south easterly direction and where the influence of severely inclement weather conditions could conceivably result in the massive movement of waterborne sediments into the newly excavated parts of the outer channel.

DREDGING FOR TRIDENT CLASS SUBMARINES

(e) The high level of liquidated damages for late completion, the cost of mobilisation of additional plant to meet acceleration requirements and the impact on professional reputation in the event of late completion.

PLANT

22. Plant comprised two main categories (as listed below) providing dedicated plant for the execution of the works and a comprehensive back-up fleet which together constituted a virtually infallible range of equipment to ensure the timely completion of the works. The freedom to change, withdraw, or increase the dedicated plant to meet any situation as it developed was maintained by the joint venture (with the agreement of ABP) but in practice, no changes occurred to the major plant items and little in respect of the smaller plant items.

23. <u>Major plant</u>:

Cutter suction dredger	"Leonardo da Vinci"
Trailer dredger	"Argonaut"
Trailer dredger	"Coronaut"
Trailer dredger	"Amerigo Vespucci"
Bucket dredger	"Mersey"
Grab dredger	"WD Dredgewell"

24. <u>Major back-up plant</u>:

Cutter suction dredger	"Taurus"
Trailer dredger	"Alfa B"
Trailer dredger	"Johanna Jacoba"
Trailer dredger	"WD Medway II"
Trailer dredger	"Cornelia"
Trailer dredger	"James Ensor"
Trailer dredger	"Galilei"
Trailer dredger	"Delta Queen"
Trailer dredger	"Flevo"
Bucket dredger	"'s-Gravenhage"

25. <u>Sea-going hoppers</u>:

"Verrazzano"
"Magellano"
"Geelvinck"
"WD Tyne" & "WD Clyde"
'Noviomagum 669' & 670'

26. <u>Back-up sea-going hoppers</u>:

"WD Itchen" & "WD Test"
"Sea Reach"
"Nyptang"
"Weseltje"

27. <u>Auxiliary plant</u>:

Multi pontoon	'DN30'
Survey tugs	'DN57' & 'DN56'
Tugs	"Michel" & 'F49'
Launch	"Johanna"
Bunker barge	'DN21'

28. <u>Electronic control system</u>   A comprehensive range of electronic position fixing equipment was provided ashore and on all vessels to facilitate accurate positioning of dredging and survey plant, dredge depth control and hydrographic survey, and record of dumps at the sea deposit site.

PROGRAMME

29. Although the original start date of January 1991 was deferred with corresponding movement of completion date, it remained the intention of the joint venture to complete the dredging works by the end of 1991 to minimise exposure to inclement weather early in 1992.   Pre-dredge survey was commenced in March 1991 (prior to formal award of contract) and the mechanical dredgers and trailer suction dredgers arrived on site and commenced work immediately in April 1991. The cutter suction dredger began the outer channel work in September 1991, primarily to minimise the impact of interim siltation on the completed sections of channel.   The dredging plant over-ran its self-imposed programme time minimally and achieved completion in January 1992 with all the works being completed comfortably within the contractual time limit. The self-imposed target completion of December 1991 was influenced by high wear rates and inclement weather affecting the outer channel work and by several marine groundings on the upper slopes, two marine accidents, and the encountering and removal of an uncharted wreck, affecting the inner channel and navigation system work.

ELECTRONIC CONTROL DETAILS AND HYDROGRAPHIC SURVEY
<u>General</u>

30. A fully equipped high speed custom built survey vessel suitable for inshore and off-shore work was employed to carry out pre- and post dredge surveys and frequent interim monitoring surveys.   Horizontal positioning was provided by Motorola Falcon 484 interfaced to an HP9920 computer spread. Nine shore stations were erected at suitable locations to give adequate geometry over the entire dredging/survey/sea deposit site areas.   The exact positions of the stations were established by triangulation from known and accepted trig points.   Vertical control was provided by two automatic tide gauges situated near Ramsden Dock Entrance and Haws Point, later substantiated by a third gauge at Halfway Shoals.   Tide readings were continuously transmitted to a receiver on the survey vessel.   Tidal information was relative to chart datum which at Ramsden Dock Entrance is 4.75 metres below ordnance datum (Newlyn).   Depths were measured by a Deso 20 echo sounder interfaced to the computer spread.   Minimum data logging of depth and position (x, y, z) was at one second intervals.   All survey data received by the computer was stored on disc.   On completion of surveys the disc was placed in a similar computer spread ashore enabling drawings to be printed using an A1 plotter.   The echo sounder was calibrated during each survey.   Calibration of the positioning equipment was carried out at regular intervals during the course of the works.   Motorola positioning

equipment and tide gauge receiver were also installed on all main dredging vessels.

Extent of surveys

31. Pre- and post dredge surveys were carried out by running lines at right angles to the centre line of the channel with a line spacing of 20 metres. The area to be surveyed covered the whole of the dredge area together with sufficient lateral over-run to provide adequate monitoring information on slope stability. At the seaward end of the channel the survey continued until the 9.00 metres below chart datum contour was reached. Very frequent interim surveys over areas required by the dredging spread were carried out during the course of the contract.

Survey reporting

32. The pre- and post dredge survey was plotted on drawings to an agreed scale. All soundings were reduced and plotted in metres and tenths of metres below chart datum. Relevant echo rolls and data sheets were provided to the engineer together with the drawings, for approval.

Positioning at sea deposit site

33. To achieve the position fixing requirements of vessels, all vessels discharging dredged material at the sea deposit site were fitted with Motorola position fixing system and computer spread. The system operated on the basis of ranges being measured between the vessel and shore stations which were established by triangulation from known and accepted trig points. These ranges were continually updated and displayed on a console situated on the bridge of the vessel. As well as being displayed, the ranges were fed into an HP9920 computer via an interface. Data including position of shore stations and position of deposit site was permanently stored in the computer which enabled the ranges to be converted to grid positions. The position of the vessel could then be displayed on a screen along with an outline of the deposit site. Because the ranges were being continually updated, the vessel's position was in turn updated on the screen. The deposit site shown on the screen had a grid superimposed over it with each square being identified by a number. The vessels were instructed to dump a specific number of loads within each square, enabling a systematic dumping programme to be carried out. Whilst this system ensured that material was spread more evenly than dumping in a random fashion, it did not eliminate the possibility of individual loads being dumped in the same position. As each vessel executed its dumping operation, the computer was instructed by a press of a button to print out a copy of the graphics shown on the screen. The printout also included the position in numerical terms, i.e. Eastings and Northings. Copies of these printouts were handed to the engineer on a regular basis.

Survey supervision

34. Survey and instrumentation was managed by a team of hydrographic surveyors and electronics engineers of sufficient number and calibre to ensure the timely provision of the pre-dredge survey, the provision of the all important day by day interim monitoring surveys, and the reliability of the survey and electronic control systems.

METHOD
Hours of work

35. The principal system of work was based on working continuously over 168 hours of the week. However, the prevailing existing levels of the channel in April 1991 entailed some initial tidal work for the trailer dredgers influenced by low water periods of spring tides and some night-time restricted work for the bucket dredger as per the agreement with Barrow Council.

Working principle

36. In principle, the outer channel was dredged by cutter suction dredger with some limited trailer dredging, and the nnner channel was dredged by trailer dredgers, bucket dredger and grab dredger.

Trailer dredging

37. Trailer dredgers executed the dredging of the granular materials and the trailable cohesive soils to the limit of their economic capacity. Two of the dedicated trailer dredgers began work in April 1991 with the initial task of reducing all existing shallow areas to the level of 3.00 metres below chart datum. These shallows existed intermittently between the area of Head Scar and seaward of Halfway Shoals and their removal was particularly required so as to provide navigational water depth at all states of the tidal period for the hopper barges working with the mechanical dredgers. Subsequent to high spot removal and concurrent with the work of the mechanical dredgers, the trailer dredgers excavated the trailable soils over the entire length of the inner channel. Tidal delays were generally avoided by reverting to naturally deeper sections of the channel during the lower tidal range, and by concentrating on the shallower areas and on the side slopes during the upper tidal range. As the works proceeded, the influence of tidal levels became less significant with increasing dredged water depths. The third of the dedicated trailers mobilised and began work in September 1991 and was engaged primarily in the outer channel in support of the heavy cutter suction dredger. Whereas the dredging of the main trailable soils was executed comfortably within the target completion date (31 December 1991) and one of the two dedicated inner channel trailers demobilised in the autumn of 1991, the other vessel remained on site until the conclusion of dredging to dredge back-well losses from the bucket dredger and to continuously maintain the toelines of the completed sections of channel and the adverse influences of

interim siltation. In accordance with the normal technique of dredging by trailer entailing the gradual and systematic lowering of the seabed, the varying nature of the dredged materials, the influences of tide and weather and the assistance in facilitating the optimum operation of the other dredging plant, the dedicated inner channel trailers were not essentially confined to localised areas and ranged continuously between areas seawards of Halfway Shoals and Ramsden Dock Entrance, and contributed substantially to the timely completion of the works.

Bucket dredging

38. Bucket dredging in the cohesive soils and dense granular material, including cobbles and boulders, began very shortly after the trailer dredgers in April 1991 at the seaward end of the inner channel. The bucket dredger worked from offshore to inshore, leaving full design level and benched lower slopes, terminating her main works adjacent to Buoy No. 7 and thereafter worked in isolated areas of the upper inner channel in areas where the grab dredger had difficulty in achieving design level. Noise restricted work system was practised in the upper inner channel and some tidal delays occurred over spring tide low waters. The Morecambe Bay gas pipe crossed the inner channel on the downstream side of Buoy No. 3 on its final landfall route to Westfield Point, thus lying in the working area of the bucket dredger and close to the relevant design level. Fully accurate information on the horizontal position of the gas pipe and its precise elevation was not available and seismic investigation was executed in an endeavour to verify top of pipe levels. It was determined that top of pipe was a minimum of 1.90 metres deeper than the relevant channel design level which represented a viable safety margin for controlled dredging by bucket dredger. Work in this critical area was very accurately monitored to an operating procedure agreed with British Gas entailing minimal dredged tolerance, and satisfactory results were achieved. The bucket dredger continued working until completion of dredging in January 1992 and her heavy dredging capability proved essential in the removal of isolated areas of hard clays encountered above the design levels in the inner channel.

Grab dredging

39. The grab dredger also began work in April 1991 in the area of Buoy No. 3 whilst the trailer dredgers were engaged in reducing the prevailing shallow depth of the upper inner channel to facilitate the subsequent continuous work of the grab dredger in this area. It may be noted that whereas the grab dredger is a shallow draughted vessel, the practical and economical loading of her hopper barges requires their positioning ahead of the dredged cut, and the pre-dredged water depths of the upper inner channel required initial deepening by the trailers, working on a tidal basis, to eliminate the tidal delays that would have influenced the grab works.

The particular advantage of the grab dredger's shallow draught was in the bench cut dredging of the upper inner channel side slopes where the dredger's 17.00 metre working radius permitted extensive lateral dredging with the attendant hopper barge lying on the channel side of the toe-line. The particular features of the grab dredger enabled her to fulfil an imperative part of the total fleet requirement. However, although she has achieved economical productions in clays with values up to 150 kilopascals, she was unable to make economical progress to design level in isolated areas of her work and recourse to final dredging by bucket dredger was necessary (as noted above).

The outer channel
40. With the exception of trailer dredging in suitable granular soils by the third of the dedicated trailer dredgers, the whole of the outer channel works was executed by the powerful cutter suction dredger equipped with barge loading chutes and supported by 1,800m³ hopper barges. The start date of September 1991 was determined as providing sufficient time for the execution of the calculated volume of work but avoiding the unnecessary exposure of completed sections of the outer channel to the influences of on-going sedimentation and possible storm damage. Cutter dredging was executed from south to north (working shorewards) covering an outer channel length of approximately 5,000 metres working at design depths, and with computerised automatic control of ladder settings for slope gradients. The high production capability of the cutter suction dredger (supported by the third trailer dredger) proved necessary to complete the works in January 1992 owing to the influence of inclement weather during the winter period, and particularly attributable to the joint venture's probable underestimation of the nature of the dredged material which included hard to very hard clays with frequent intrusions of very dense granular deposits and the incidence of random boulders extending to concentrations of boulders. Boulder occurrence required that the range of overdredging was extended in limited instances to levels as deep as 2.00 metres below design level. It is very unlikely that the employment of any alternative dredger to "Leonardo da Vinci" would have permitted the timely execution of the works.

Reserve Plant
41. It was not essentially within the contemplation of the joint venture that replacement or additional plant would be required to assist in the main excavation works. However, a massive hydraulic dredging capability was always available in reserve to cater for winter sedimentation damage to the dredged outer channel, and the adverse influences of side slope instability, both of which were areas of extreme risk.

Other Associated Works
42. Construction of leading light structures and beacons began in April 1991 with tidal zone construction being

executed by land based plant using temporary haul roads and offshore construction being executed from floating plant. One structure suffered damage, due to marine accident involving one of the joint venture's seagoing hoppers, and was subsequently repaired and all the works were satisfactorily completed within the required programme time. The inter-tidal zone haul roads and the burying of the electric cables providing mains power were executed in accordance with a conservation programme agreed with Nature Conservancy Council and final reinstatement completely eliminated minor temporary damage to the extensive areas of littoral flora in the confines of Sites of Special Scientific Interest. It should be noted that Nature Conservancy Council (now English Nature) extended a very realistic and active co-operation to the sub-contractor in his execution of the navigation system works.

MANAGEMENT

43. The joint venture was constituted as an autonomous organisation for the execution of the works with parent company guarantees between the constituent members and with individual parent company guarantees to ABP for the entirety of the obligations undertaken by the joint venture. The joint venture was governed by a board of three directors with each parent company providing one director. The joint venture was managed by one project manager assisted by three operational managers drawn from the three constituent member companies with each operational manager being responsible for the plant provided by his company and being responsible for the work allocated to his plant by the project manager. Joint venture board meetings were held on site at Barrow at frequent intervals with the project manager only holding responsibility for the overall execution of the works and reporting to the Board of Directors. The on-site management was assisted by a sufficient number of dredging supervisory personnel drawn from the constituent member companies. In practise, the entire management and supervision worked with good and businesslike discipline and with harmonious accord as exclusive employees of the joint venture.

SAFETY

44. The confined nature of the inner channel in particular required first class professional seamanship and close co-operation with Barrow Harbour Master's Department. The marine accidents and groundings referred to above were essentially due to errors of judgement and did not result in serious loss or irremediable situation.

QUALITY ASSURANCE

45. The joint venture submitted with its tender a preliminary Quality Assurance Plan and comprehensive Safety Manuals. On award of contract, the joint venture appointed its own quality assurance manager who worked in close co-operation with ABP in the preparation of fully detailed Quality Assurance Plan in accordance with AQAP(NATO) and

Defstans requirements. Quality assurance formed an important aspect of the control of the works and regular quality audits were initiated at the direction of ABP.

CONCLUDING COMMENTS
Major risks
46. The impact of inclement weather proved more onerous than the pre-tender assessment but without impediment (in isolation) to the target completion date. The majority of the material experienced in the outer channel, and to a lesser but significant extent in the inner channel, proved more onerous than the pre-tender assessment but without serious impediment (in isolation) to the target completion date. Both the above had detrimental impact on the average levels of overdredging for the cutter suction dredger working in the outer channel and, in respect of material, for the bucket dredger working in the inner channel. Side slope final gradients did not vary significantly from the design gradients and instability of the slopes did not materialise. The movement of finer particles down the slopes to their cumulating position on the toelines and extending into the fairway required the constant application of a trailer dredger during the final stages of the works, but this remained generally within the extent of expectations. Similarly, interim siltation entailed some additional trailer work generally within the anticipated range. The incurring of liquidated damages was never anticipated by the joint venture and remained inapplicable. The minimal over-run of the self imposed completion date was attributable to inclement weather during the winter period of 1991-1992 and to the nature of the material that was dredged (taken cumulatively rather than in isolation).

Quantities
47. Volume to be dredged to design level was approximately $3,000,000m^3$. This volume had limited relevance to the joint venture which had the obligation to deliver the designed works irrespective of net volume, and to make its own assessment of overdredge, slope stability, interim sedimentation and storm damage influences. The gross volume finally dredged was approximately $4,250,000m^3$ exclusive of on-going siltation which was difficult, if not impossible, to quantify. However, in recognition of the occurrence of interim siltation it is not unlikely that the final gross volume may be rounded off to some $4,500,000m^3$.
The outer channel involved marginally more than 50 per cent of the final gross volume.

Wear and tear
48. Wear and tear resultant from dredged material was higher than anticipated for all the major plant items and substantially higher than anticipated for the cutter suction dredger working in the outer channel.

Conclusion

49. Prudent selection of plant by the joint venture and the noteworthy harmonious and professional dedication of all the on-site personnel, provided for the timely completion of the works and the fulfilment of all contractual obligations to the complete satisfaction of Associated British Ports.

Coulport and Faslane general management

K. PARTINGTON, Project Director, TBV Consult, CSB Faslane

SYNOPSIS. The Government's decision in 1980 to replace the Polaris weapon system with Trident entailed a huge redevelopment of the Clyde Submarine Base at Faslane and RNAD Coulport. PSA Projects (now called TBV Consult, part of TBV Professional Services) managed the £1.7bn development - one of the biggest and most complex projects in Europe - on behalf of the Ministry of Defence. Considerable environmental, technological and political problems had to be overcome. Many facilities were designed, built and commissioned to rigorous nuclear safety standards, developed with the safety authorities during the course of the project. The development was successfully completed in 1993.

INTRODUCTION

1. The Property Services Agency (PSA) carried out the outline planning for - and then project-managed - the £1.7bn development of the Clyde Submarine Base for Trident, on behalf of the Ministry of Defence, from 1980 through planning and construction to completion in 1993. This period saw changes in the relationship between the two departments which are outside the scope of this paper, culminating in PSA's entering the private sector in December 1992 as PSA Projects Ltd. It is now known as TBV Consult, part of TBV Professional Services. However, to avoid confusion, this paper describing the management arrangements for the development uses the name 'PSA' throughout.

BACKGROUND

2. In 1980 the Government decided to buy from the United States the Trident weapons system and supporting components for a force of British missile-launching submarines to replace Polaris. The submarines would be based at the Clyde Submarine Base at Faslane, and would rely for missile storage and servicing on new facilities to be provided on a site adjacent to the Royal Naval Armament Depot at Coulport.

3. In September 1982 the Government decided that the United States should undertake the initial assembly and long-term refurbishment of British Trident missiles at Kings Bay in Georgia. This led to a reassessment of the facilities required at Coulport.

CONTENT OF PROGRAMME

4. The shore support facilities required for Trident and other submarines comprised some 100 projects, ranging from simple stores buildings to the huge shiplift and the floating explosives handling jetty. Between these two extremes, the programme included a 2-berth finger jetty, the modernisation of berths, new power stations and utilities buildings, reinforced concrete explosives handling and storage buildings, workshops, laboratories and training buildings, land reclamation and dredging, and all the necessary infrastructure of roads, services, drainage, security fences and systems and landscaping. It was by far the largest and most complex development ever to be tackled by PSA and MOD. At its peak it was second only to the Channel Tunnel project. For a time, political opposition by local authorities posed problems. However, it was the tightness of the dates by which the MOD required the key support facilities, and the need to demonstrate that designs would meet the stringent requirements of the nuclear safety authorities, that would present the project team with its greatest challenges.

ENVIRONMENTAL IMPACT ASSESSMENT (EIA)

5. From the outset PSA gave full consideration to the potential impact of the new development on an area of natural beauty, and on local communities, and particularly the volume of construction traffic it would put on local roads.

6. In May 1984, PSA produced an Environmental Impact Assessment (EIA) (one of the first in the UK) which addressed all the environmental issues in detail, with proposals to alleviate the impacts wherever possible. These included the construction of new public and private roads, which remain of permanent benefit to the local area; and detailed design and landscaping proposals, subsequently incorporated in a design strategy document agreed with the Royal Fine Art Commission for Scotland and the Countryside Commission for Scotland. The EIA also foresaw benefits: the development would inject substantial additional monies into the district and regional economies, through the creation of new jobs (in the peak years more than 3000 construction jobs were directly created) and the purchase of materials and services.

7. These proposals were presented to local communities by a MOD/PSA team in a series of public meetings in 1984. (Further public presentations were given to local communities in 1987 shortly before the major contracts started).

8. An outline Notice of Proposed Development was submitted to Dumbarton District Council with the EIA, but it had to be referred to the Secretary of State for Scotland for decision.

9. In giving outline planning consent in March 1985, and after consultation, he endorsed, inter alia, the proposals for new roads. He also imposed a requirement to bring in the bulkiest and heaviest loads by sea, including the floating jetty at Coulport and the shiplift platform and piles at Faslane, and virtually all concrete making materials by sea or rail.

THE CONSTRUCTION SITES

10. The construction work was concentrated on 2 main sites, at Faslane and Coulport. A third site was opened in 1988 at Hunterston near Largs, for the construction of the floating jetty - whence it would be towed to its permanent mooring at Coulport.

11. The majority of the Coulport facilities were provided on a 235 hectare greenfield site on the Rosneath peninsula, an undulating area of peat and woodlands sloping steeply into Loch Long on the west and presenting very difficult ground conditions. A strategy for peat clearance and disposal, in natural hollows where possible but sometimes with retaining bunds, was developed in the EIA and was successfully carried out. Even so, deep peat sometimes caused problems, such as the need to pile the foundations for a section of security fence.

12. There was plenty of rock, too, though being a mica schist it degraded quickly and often required much greater excavation or special pinning back. Some 500,000 m^3 of rock spoil was used as a screen to improve the visual impact of the depot as seen from Ardentinny. Some works also had to be carried out in the operational depot. The interfaces there with existing services and operational requirements created significant challenges for the project team and the depot.

13. The Faslane works were in 2 main areas. One lay in the heart of the congested, busy and highly security-conscious naval base. The other, the 24 hectare northern development area, was effectively a greenfield site. However, it had previously been leased by shipbreakers. Such activities produce asbestos, but a dump had been identified to MOD and PSA and had been cordoned off. However, a survey in mid-1984 revealed that most of the site was contaminated with asbestos. This entailed a huge decontamination operation, with strict air monitoring and other environmental controls. Off-site disposal in a licensed tip was ruled out because of the potential impact on local communities. However, with the specialist expertise of Balfours Consulting Engineers, all the contaminated material was either treated or safely protected and contained on site very successfully. Much of the raw asbestos was vitrified, the first use of this method on such a scale. The clean site was regularly inspected by the Health and Safety

Executive (H&SE) and was approved by them for further development.

14. It was one of the largest asbestos protection and monitoring sites at the time and established new standards and criteria, agreed with the H&SE, for this type of hazard. It required careful management to meet this extra challenge and maintain the Trident works programme.

15. Besides land sites there were works in the Gareloch, where PSA's dredging activities - in Faslane Bay in the area of the proposed shiplift and at Rhu Narrows to provide deeper and wider access for the Vanguard submarines - attracted keen local and parliamentary interest. Both operations entailed consultations with licensing authorities, conservation bodies, fisherman's associations and others. In Faslane Bay, the dredging of some 65,000 m^3 of silt and boulder clay was complicated by a layer of saturated asbestos on the sea bed. Great care was taken to ensure satisfactory environmental controls on the dredging operation, disposal in an underwater dumping ground in the Firth of Clyde and subsequent monitoring of turbidity, suspended solids and dissolved oxygen. The Rhu Narrows dredging entailed the removal of some 2 million m^3 of material including about one-third of Rhu Spit, a natural feature popular with local people especially anglers. There was also local concern, partly allayed by advice from the Hydraulic Research Laboratory, about the possible effects of the dredging on navigation, moorings and fishing.

16. In the event, both the dredging operations were carried out very successfully without any adverse after effects on the local environment.

THE PROJECT MANAGEMENT TEAM

17. The range of management issues faced by this development - technical, financial, political, legal and logistical - was formidable : many inter-related, some conflicting. The main ones were:

> (a) Financial: control and reporting of works costs and consultants' and in-house fees and charges; Vote management and accountability, through PSA's Chief Executive, to parliament (financial responsibility and accountability transferred to MOD from 1 April 1990 as part of the process of 'untying'); also pre - 1.4.90, exercising delegated financial responsibility from Treasury and seeking fresh approvals.
> (b) Environmental - and often politically sensitive - issues: asbestos; dredging; construction traffic and noise; outline and detailed planning approvals; visual impacts; landscaping. All had to be managed against a background of parliamentary interest in the Trident programme (while a government agency PSA reported to Ministers and supported them in responding to matters raised in the House, by constituency MPs etc) and local authorities opposed to the development.

(c) Planning, programmes and logistics.

(d) Legal - eg a legal challenge on the rights of the Crown to obstruct a public highway.

(e) Consultants: briefs, awarding of commissions, payment of fees, management and integration of designs. Some 450 sub-commissions were placed with 71 main consultants.

(f) Design standards to meet the stringent requirements for nuclear installations.

(g) Rigorous quality assurance for design, construction and project management.

(h) The nuclear safety case.

(i) Contract strategy: tendering, awarding contracts, site supervision organisations.

(j) Management of the phased handover of facilities, including clearance of snagging and inter-facing with the base and depot with regard to long-term maintenance.

18. All these and many others had to be controlled within a tight programme with a very high level of inter-dependencies.

19. In view of the size and complexity of the development, PSA created in 1983 a new directorate of professional staff in Croydon.

20. It had to be expanded over the next few years as the nuclear safety and QA requirements became clearer. A joint MOD/PSA planning team was set up at Faslane to interface with the Croydon team, to draw up a development plan leading to the final layout of the facilities in relation to all the local and national planning constraints, to assist in writing briefs, to plan the integration of services across the entire submarine base and to plan the layout and support facilities across the construction sites.

21. PSA recognised that once work started on the ground, an equivalent team structure would be needed on site for construction management. This process started when the first contracts were awarded in 1985. Over the next 2 years, as further contracts were let, the team and the work shifted almost entirely from Croydon to Scotland. A PSA director was on site from 1987 when the shiplift package started.

22. During construction, the director was supported by 7 assistant directors. Four were responsible for construction works, each with a specific area: Coulport; Hunterston (the floating jetty); Faslane southern area (the existing base); and the Faslane northern development area (the NDA). Typically an assistant director managed 5 or 6 project managers each responsible for a £20m-£100m contract package. Each PM was responsible for the managerial and financial aspects of his projects to ensure delivery to time, cost and the required quality; the supervision and control of his consultant team; adherence to the brief

and a realistic programme; and the effective and economic use of resources. PSA's own systems were used for forecasting and monitoring key targets, milestones and inter-relationships between projects and between contracts.

23. A further 3 teams headed by assistant directors managed the vital support functions:

> (a) production of the design safety case; quality assurance (the directorate achieved BS5750 Pt.1 accreditation on 24 October 1991 for the project management of facilities for nuclear safety); lifetime records; operation and maintenance manuals to nuclear safety standards; testing and commissioning arrangements for nuclear related facilities.
> (b) overall management of works costs, consultants' and in-house fees and charges, claims and final accounts.
> (c) financial approvals from MOD and Treasury; ministerial and parliamentary matters; public relations; interfaces with other government departments.

24. A close working relationship between PM and client was essential to the integration of the latter's evolving requirements, especially in the field of nuclear safety. PSA and MOD (Navy) had worked closely together on many previous occasions.

25. Early in 1985 MOD appointed their own Trident works project manager to provide a single point of coordinated advice on all client matters - unusual at that time within government, though common practice now. A strong committee and reporting structure was established to ensure regular reviews of progress against time and cost and to highlight potential problem areas. Respective MOD and PSA responsibilities for the projects in construction at the Clyde Submarine Base were formally defined in a joint statement of understanding. Detailed monthly reports were provided to MOD on the cost of the projects throughout construction. With the advent of client ownership, MOD became more closely involved in the general management, culminating in the appointment of their own Director of Works in April 1992.

26. In the later years of the development, PSA's preparation for commercial accounting and privatisation generated a number of improvements in their financial systems.

CONSULTANTS

27. It was government policy that most of the design work, and all the construction work, should be carried out by the private sector. In selecting consultants, PSA were naturally looking for firms with a high level of professional experience, knowledge and skill. Because of the

location of the development and whenever it was reasonable to do so, selection was deliberately biased in favour of Scottish firms or those with an established office in Scotland. Competitive bidding was not used at that time. Selection was by interview, with negotiation of fee appropriate to the size of the job.

28. Non-standard partial commissions were generally used, based on PSA/ACE Standard Conditions or RIBA as appropriate. These were usually based on PSA plan of work stages 1-4: feasibility, outline to detail design, construction and post-construction; they included additional duties as necessary. Most commissions included construction control, but not site control for which separate commissions were awarded to consultants responsible for design.

29. Multi-discipline teams, comprising architects, civil engineers, M&E engineers and QSs, were established for specific parts of the development: for the southern area at Faslane; for the shiplift, utilities building, finger jetty and cranes in the northern area; for the floating jetty; and for the other packages of work at Coulport.

30. In each case one architect was appointed to coordinate building designs across that part of the development in accordance with the EIA and the approved design strategy.

31. Design lead consultants were appointed to coordinate design, their discipline being related to the predominant nature of the works. Designated lead consultants were responsible for the coordination of site works. The PM was the key point of contact in each case. However, a PSA liaison officer was appointed for the general management of each commission and to be responsible for the payment of fees etc (but not to provide design input).

32. The consultant quantity surveyor was responsible to the PM for cost control during design and for cost management during construction.

FAST-TRACK PLANNING

33. Ideally, every job would have been fully pre-planned before the start of construction. In practice, during the design stages some two-thirds of the work by value had to be put into continuous planning in order to hold the programme. This "fast-track" approach meant that the normal breaks in the design process for scrutiny and discussion of the design and costs with the client at completion of each design stage had to be sacrificed; the scrutiny and decision-taking process still had to proceed, but in parallel with continuing design on the next stage. PSA and MOD recognised the risks, but it was essential. Special steps were taken to maintain cost control: for example, joint PSA/MOD cost and design audits of key projects. However, the development of the nuclear safety requirements and the clarification of client requirements for the more technically complex facilities increased the design task. In the circumstances, PSA frequently had to invite tenders and even start

construction before design was complete.

NUCLEAR SAFETY

34. PSA were required to produce a design safety case, ensuring that the designs of all new land-based structures, systems (eg power and cooling water) and equipment (eg cranes) in support of nuclear propulsion plants were justified against rigorous safety criteria. This was new ground for PSA and for many of the consultants: PSA took their full share of the limited pool of expertise.

35. At the outset, however, the safety criteria were not fully defined in the briefs to PSA. The complexity and magnitude of the safety case became clearer as designs were developed and as a firm interpretation of the nuclear safety authorities' standards was obtained from their comments on the emerging safety cases. PSA had to control this process and manage the effects on designs, programmes and costs.

36. All nuclear safety-related facilities and services had to be shown to be safe in the event of a 1 in 10,000 year seismic event, with a reference earthquake at 0.2g in addition to other criteria. The generation of the safety case documents was a complex and very onerous iterative process involving design consultants and PSA's specialist consultant safety advisers, NNC Limited.

37. A safety case taking account of all construction changes will be in place for operational facilities in 1994, as required, and achieved through a fully auditable QA regime to BS5750 and BS5882.

38. PSA operated its own QA system and closely monitored and controlled the QA procedures of all organisations in the project to the required standards - a very resource intensive process.
A strict regime of mechanical and electrical commissioning and testing was imposed to ensure that the nuclear safety requirements had been satisfied. A system of independent qualified test teams was set up, covering 11 main areas. The teams were located on site, drew up the necessary specifications and checklists, and in the peak commissioning period they often worked round the clock. A Site Test Authorisation Group was set up to identify the extent of testing, authorise the methods used and ensure satisfactory results were obtained prior to handover.

39. PSA also managed the production of supporting documentation for the operation and maintenance of nuclear safety-related facilities and services at the base; and the installation of a sophisticated computerised system for controllable storage and management of records for the through life running of the base.

CONTRACT STRATEGY

40. As a government agency, PSA was normally required to adopt competitive tendering and use GC Works 1 Edition 2 - the General

Conditions of Government Contracts for Building and Civil Engineering Works. The opportunities for using design and construct were very limited, but this approach was adopted where appropriate - for example, for the specialist cranes.

41. At that time, it was PSA policy to invite and appoint nominated sub-contractors for complex mechanical and electrical works. However, there were some early problems between main and sub-contractors in agreeing contractual conditions and programmes. In order to eliminate such risks, PSA decided to switch almost exclusively to domestic sub-contractors. PSA would then deal only with the main contractor, who would be responsible for the selection (from a list approved by PSA) and appointment of his own subbies and thus be the sole point of responsibility. This was a major departure from contemporary government practice and a step forward to achieving targets.

42. The contract strategy was based on a number of considerations: the integration of new facilities with the existing base and depot, and with each other; grouping of facilities of a similar type if their programmes matched and they were reasonably closely located; the risk of too many contractors working in one area leading to contractual disputes and claims; and the undesirability of too many key projects in the hands of one contractor, though subsequent takeovers negated this to some extent.

43. Accordingly the projects or briefs at the CSB were arranged into more than 70 main contracts. Some briefs - for example, for services across large areas - were split up between a number of contracts. Each project was developed and costed on an individual basis up to pre-tender stage; from that point costs had to be reported for contractual reasons on a contract package basis. But PSA also reported to MOD and Treasury the costs of the individual elements of each package.

44. The contract strategy envisaged the use of advance works to prepare and service sites and so to minimise the risks to subsequent major contract packages. PSA prepared from the outset a major package of advance works to delineate and open up the new Coulport site - peat clearance, roads, security fences, drainage etc. The strategy also provided the flexibility to respond to changing needs. For example, it was later decided to add a further package of roads, security fences and excavation of foundations, thus reducing the volume of the major B&CE works package for the new explosives area, and also making good use of a sitting contractor with all the appropriate equipment already on site. At Faslane, another example, a package of enabling works - drainage, site roads and site levelling - was negotiated into the asbestos clearance contract, thus ensuring a rapid start up for the critically important shiplift package.

CONSTRUCTION PROGRAMME COORDINATOR (CPC)

45. Following consultation with the construction industry, PSA concluded that the management organisation on site should be strengthened by the appointment of a major contractor. His construction experience would better equip him to coordinate such a huge programme with the interfaces, requirements and tight timetables for so many different contractors. He would also be able to provide specialist skills - for example, in industrial relations and planning.

46. Following competitive tendering, Wimpey Major Projects were appointed as construction programme coordinator (CPC) in 1987. The CPC was supervising officer (SO) on most of the contract packages from that date. The SO had delegated responsibility for the construction of each contract to time and budget, reporting directly to the individual PMs.

CONSTRUCTION NOISE

47. The EIA had recognised the potential impact on local communities of construction noise and vibration, especially from the very intensive marine piling operations in Faslane Bay, and identified measures to reduce this. With the advice of specialist consultants, W S Atkins, a strategy was devised to control on-site noise to meet the actual constraints imposed by the Secretary of State for Scotland. It included the erection of fixed noise monitoring stations at key locations, clear guidelines to contractors (which for the most part were observed admirably) and close monitoring on-site with hand-held equipment. For people living outside the sites but within certain defined levels of construction noise, MOD provided double glazing, and a handful of properties had to be purchased. The dredger working in the Rhu Narrows, close to a number of dwellings, was fitted with acoustic baffles.

48. Naturally, local people complained when noise seemed excessive and Dumbarton District Council's environmental health officer rightly kept a close eye on this aspect of the development. But in fact remarkably few complaints were received - and when they were, prompt and effective action was taken. What had threatened to be a serious potential source of annoyance - and delay if work had to be stopped - proved to be entirely manageable by good practice.

TRAFFIC MANAGEMENT & DELIVERY OF BULK MATERIALS

49. Good management of construction traffic was always of key importance. When the Secretary of State for Scotland gave outline planning consent for the development in March 1985, he made it a condition that the 2 new roads identified in the EIA -the public Garelochhead bypass and the MOD's private (but open to the public)

northern access road - should be built as quickly as possible.

50. The EIA had concluded that there was no opportunity for the early construction of a bypass around Helensburgh and Rhu. However, the Secretary of State for Scotland asked that some further thought be given to ways of relieving Helensburgh and other local communities from the burden of traffic. PSA and MOD subsequently decided that this could best be achieved by building a temporary road specifically for construction traffic between the A82 on Loch Lomondside south of Luss and the new bypass. The latter was by then already under construction. The 14km Glen Fruin road was built in under a year - and well inside its budget - and all 3 roads were opened at the beginning of 1988, just in time for the real build-up of construction traffic. Prior to their opening, a preferred route was designated to which all contractors' vehicles had to adhere; and lorries were banned from passing through Helensburgh and Garelochhead when children were going to and from school. Once open, the new roads became the designated route and construction traffic ceased to be a real issue.

51. The Glen Fruin road was intended for the construction traffic - at peak some 2,500 vehicle movements a day - generated by a multiplicity of contractors and sub-contractors using material and labour from many different sources. Clearly road transport was the only sensible answer. Bulky and very heavy construction items were a different matter. In view of the earlier undertaking that the bulkiest and heaviest items would be brought in by sea or rail, PSA and MOD had examined very carefully the alternative options. The sea route was the natural choice since the Coulport and Faslane sites were directly accessible by sea: the facilities were available at negligible cost and with minimal arrangement, whereas the rail network did not, and could not entirely, reach both sites.

52. A temporary jetty had in any case been provided at Coulport for the use of the advance works contractor. At Faslane, in order to provide containment for additional material contaminated with asbestos, the line of the new sea wall was adjusted prior to construction around the old "beaching dock", (formerly used for breaking up battleships), thus also providing a natural berth for the delivery of bulk materials directly into the northern development area. Concrete, roadmaking materials, armourstone etc brought in by sea to Coulport totalled some 800,000 tonnes; at Faslane, concreting materials, piles, steelwork, reinforcement, pre-cast concrete units etc brought in by sea, totalled some 350,000 tonnes.

53. In addition, the 85,000 tonnes floating jetty was towed by sea from Hunterston to Coulport; the shiplift platform, and the 125te and two 20 tonne dockside cranes were all brought into Faslane by sea.

INDUSTRIAL RELATIONS

54. Following discussions between PSA and the Federation of Civil Engineering Contractors, it was agreed early on that nominated site status would not be appropriate because, taking the programme as a whole, the B&CE works were significantly greater in proportion to the M&E works and also the works were at 3 quite distinct locations. Although there were some relatively minor difficulties at the start of the M&E phase, the industrial relations strategy subsequently adopted by PSA on the advice of Wimpey, to ensure that compatible industrial relations policies were used by all the contractors, was highly successful. This was achieved by a two tier approach: a management group of all main contractors and their sub-contractors, under the chairmanship of Wimpey; and a lower level contractors' group to implement the management group's strategy and deal with site issues. Contractors maintained full responsibility for their workforces, managing them within the agreed policies, procedures and programme. Each contractor dealt individually with the requirements of his respective industrial agreements and trade unions.

55. The success of the policy can be gauged by the fact that during the course of the development, less than 0.2% of some 3.5m man days over the 8 years of construction - a period of boom as well as recession - was lost to industrial disputes.

SITE SAFETY

56. Similarly the health and safety record for this huge development was remarkably good by any standards. There was one recorded accidental death on site - and even one is a tragedy - but that has to be seen in the context of 3.5m man days over an 8 year period. PSA's own site safety officer and CPC worked very closely with the Health and Safety Executive, and cooperation by the contractors was excellent.

CONCLUSION

57. The CSB development was immensely challenging in every way. The programme was always very tight. Considerable environmental, technological and political problems were encountered, not the least of which was nuclear safety. PSA allowed for cost and programme risks. However, the full extent of the infrastructure needed for Trident only gradually became apparent. The very nature of nuclear safety and the iterative process of building a safety case meant a continual re-appraisal of costs, programmes and management systems. PSA, their consultants and CPC reacted to control the effects of changes in safety requirements on the design and construction process. In spite of these difficulties, all the facilities were successfully completed in time to meet the operational requirements of the Trident programme.

Safety case development for Trident facilities

B. V. DAY, Head of Structural Mechanics Department, and
P. H. DAVIDSON, Group Head, Systems and Safety, NNC Limited

ABSTRACT. This paper addresses the development of the safety cases for the Trident facilities at CSB Faslane and RNAD Coulport. It explains the framework of the documentation required, the logical processes by which detailed safety requirements were derived, and the studies that were necessary to demonstrate that the design meets these requirements. The comprehensive and structured approach is illustrated by examples.

INTRODUCTION

1. NNC's involvement in the Trident facilities at CSB Faslane and RNAD Coulport began in 1985 when designs for the new works had already been drafted. These developments provide for the berthing, maintenance and repair of nuclear powered submarines. When berthed, the submarines have a (limited) dependence on shore services to maintain reactor safety. Additionally, the Shiplift facility is required to support a submarine stably to avoid jeopardising reactor safety.

2. Initially NNC were approached by PSA to assist in the preparation of a structural assessment of the Shiplift platform at Faslane. From this NNC's role grew to include all the new facilities, viz the Shiplift, Finger Jetty, Northern Utilities Building and Service Infrastructure at Faslane, and the Generating Station, Floating Jetty and Service Infrastructure at Coulport.

3. NNC's responsibilities encompassed:
 - the production and presentation of a structured case for nuclear (reactor) safety
 - reviews of design changes and deviations for their impact on the safety case
 - safety input and advice to PSA and their consultants.

4. This paper reviews the purpose of the safety cases and explains how they were developed for these facilities. The systematic and structured development of the safety cases is demonstrated by an example. A second example then illustrates the approach to the detailed analysis of the Shiplift platform, in support of the safety case.

SAFETY CASE STRUCTURE

5. The purpose of a safety case is to act as the vehicle for obtaining safety approval of a design. The key to constructing a safety case is a knowledge and understanding of the overall safety requirements. For the Trident facilities these stem from the submarine reactor systems which for security

SAFETY CASE DEVELOPMENT

reasons had to be treated as black boxes. NNC therefore sought guidance on safety requirements from the MOD and their safety advisors. Initially the only established safety requirements covered seismic capability and segregation against localised hazards (e.g. fire) to be applied to all the new facilities. However by June 1986 a provisional set of safety requirements was established which applied to:
- overside services to submarines
- the Shiplift as a structure/mechanism supporting a submarine

6. A safety case is a structured suite of documents which provides a detailed derivation of the safety requirements for the facility systems and structures. The case is completed by the supporting substantiation of the design to demonstrate that the requirements are met.

7. For these particular projects it was clear that the safety case structure would have to allow for an initial fluidity in the definition of the overall safety requirements and the multiplicity of design consultants and contractors.

8. The resulting structure is shown, for a typical facility, in Fig 1. The pyramid structure is headed by a Design Concept Document (DCD), supported by System Functional Specifications (SFSs), which in turn refer to Technical Reports (TRs).

9. Part A   Design Concept Document

The function of this report, compiled by NNC, is to give an overview of safety for the whole facility. It states the fundamental criteria which the case is to address and outlines the concepts and principles whereby the facility meets these criteria.

10. Part B   System Functional Specifications

These are functional statements for the whole facility, divided into separate documents each covering a separate engineering discipline. They were all produced by NNC. The function of each document is to define in a systematic manner all of the safety duties and requirements of each system, plant item or structure which is covered by its scope. Having outlined the design, the document then demonstrates how the design meets the identified duties and requirements, referring to reports from the further group in Part C, as necessary.

11. Part C   Technical Reports

Part C reports substantiate details of the safety case presented but not proved in the Parts A and B documents. These reports represent the bulk of the safety case. They were mainly prepared by the appropriate design consultants in accordance with synopses produced by NNC.

12. The way in which the safety report package for a facility provides a structured and comprehensive safety case is best illustrated by examples. The first example explains the approach at the DCD and SFS levels. It covers the systems for controlling electrical equipment that provides overside 60Hz supplies to berthed submarines. A further example illustrates the role of the TRs and the level of detail provided in them, by describing the detailed structural analysis of the Shiplift platform.

SAFETY CASE DEVELOPMENT - CONTROL OF ELECTRICAL SUPPLIES

13. The electrical supplies at Faslane and Coulport are under operator control. Operator action is required to control switchgear, diesel-alternators and associated plant as necessary to achieve recovery from any faults that affect the supplies at the site.

Overall safety requirement

14. The provisional safety target provided by MOD and SRD can be summarised as:

'Loss or inadequacy of overside 60Hz supplies for a period in excess of 20 minutes or 60 minutes (depending on the vessel state) should not be more likely than 10^{-4}/year'.

Design provision

15. A review of the design indicated that two separate systems are provided to enable the operators to control equipment. These comprise a computer-based system (SCADA) and a separate hardwired system.

16. Each plant item to be controlled therefore has a switch to select the control system in operation.

Consideration of localised hazards

17. A localised hazard, such as a fire (with a nominal likelihood of 10^{-2}/year), will be more likely than the target frequency for loss of overside 60Hz supplies. Therefore at least one of the two control systems must survive a localised hazard. This was most readily achieved by:
- Segregating the two control systems from each other,
- Segregating the selection switches (i.e. locating them with the plant, which is also in segregated groups for similar reasons).

18. The logic is shown in Fig 2. However locating the selection switches with the plant, in distributed segregated groups, means that, following a fault on the operating control system, it will take longer than 20 minutes to reselect to the other control system. It is therefore important that a localised hazard that damages the operating control system does not simultaneously cause an interruption in the overside 60Hz supplies. It must also be recognised that, at a given time, either of the control systems may be in operation.

Reliability considerations

19. A common cause of an interruption in the 60Hz supplies is a brief loss of grid, estimated to occur bi-annually. The design requires operator control following this and similar events, and the target reliability for the 60Hz overside services therefore requires a reliability from the control systems (and operators) of about 10^{-5} failures per demand.

20. Since it is expected to take more than 20 minutes to transfer to the alternative control system, it is clear that whichever control system is in operation at the time of the interruption should achieve the 10^{-5} failure per demand target.

21. For the same reason, it is also important that the operating control system should not emit multiple spurious tripping signals at any time. These aspects are summarised in Fig 3.

Performance aspects

22. A review of the electrical system showed that some 40 separate operator actions are required to recover from the most severe conditions causing an interruption in the 60Hz overside services. The consequences of this are shown in Fig 4.

23. Allowing a time margin for operator diagnosis and planning, and a contingency for dealing with plant faults, about 20 seconds are available for each control action. The control system response times had to be commensurate with this.

24. The operator reliability target deduced above, if achievable at all, required ergonomic aspects to be optimised.

SAFETY CASE DEVELOPMENT

Consideration of widespread hazards

26. It has already been established that:
- whichever control system is in operation at a given time must survive any credible incident that can cause an interruption in overside 60Hz supplies and therefore a need for the control function
- whichever control system is in operation must not spuriously generate multiple plant tripping signals.

26. These aspects apply, inter alia, for hazards with a widespread effect, e.g. a seismic event. The logic is shown in Fig 5.

Summary of safety requirements

27. The outcome of the foregoing considerations is shown in Fig 6. This represents the level of detail that was presented in the (top level) DCD. The information was a sufficient basis for a specialist to develop the appropriate SFS and TR requirements, taking account of any factors that may arise from the specific nature of the equipment or structure under consideration.

Interfaces

28. Interface requirements can fall into two categories:
- Interface aspects that impose requirements on the equipment or structure under consideration
- Aspects of the equipment or structure that impose requirements on other equipment or structures.

29. Using the above examples, the overall numerical reliability and response time of the control systems were derived from consideration of the electrical and mechanical equipment to be controlled. However the control systems themselves rely on uninterruptible power supplies, and on the civil works for segregation and protection against (or a known response to) the identified localised and widespread hazards.

30. The systematic thought processes that are inherent in the above steps allowed clear recognition of the interface aspects which were recorded in the SFS covering the equipment under consideration and carried across to the companion SFS that address the interfacing function.

Supporting substantiation

31. For the Northern Utilities Building, the Technical Reports required to substantiate assertions made in the SFSs are shown in Fig 1.

32. The flexibility required from the safety case structure was achieved as follows:
- Each TR was a discrete package that could be separately prepared at the optimum time for the particular equipment or structure to be covered. Generally, civil works preceded M&E and control equipment.
- As a general rule, each TR required input from only one Consultant or Contractor, and one area of specialisation within NNC

33. The above example covers only the control function. When the requirements for electrical, mechanical and structural functions in support of the overside 60Hz supplies are added, it is clear that the systematic and structured approach to the safety cases that has been described in this paper was crucial in ensuring that all the detailed requirements had been correctly identified.

34. In practice, it was necessary and useful to both NNC and the consultants and contractors for NNC to prepare synopses of the TRs, after careful consideration of the requirements identified in the SFSs.

SAFETY CASE DEVELOPMENT - STRUCTURAL ANALYSIS OF THE SHIPLIFT PLATFORM

35. This second example illustrates the way in which the detailed structural analysis of the Shiplift platform was developed and implemented by NNC in support of the overall safety case for the Shiplift Facility.

Overall structural analysis objectives

36. The objectives of the structural analysis carried out in support of the Shiplift safety case were as follows:
- To demonstrate the structural integrity of the Shiplift platform
- To substantiate the structural integrity assumptions inherent in the safety case for the Shiplift control and protection systems, which are intended to limit loads on components to safe levels.

Structural analysis strategy

37. The strategy developed by NNC to meet the above objectives comprised four phases as follows:
- Estimation of enveloping loads
- Structural integrity analysis under the estimated enveloping loads
- Global response analysis of the Shiplift platform to calculate the expected loads under a variety of loading conditions
- Reconciliation of the global analysis loads with the estimated enveloping loads.

38. This strategy represented the most efficient approach with regard to both cost and programme. It enabled the detailed design and safety case to proceed in parallel with the minimum of risk. This was achieved by performing the structural integrity assessment early in the work programme in parallel with the global response analysis so as to enable any potential design modifications due to the safety case requirements to be easily incorporated before finalisation of the design.

39. A brief outline of these phases is given below.

40. *Estimation of enveloping loads* Enveloping loads for the major components in the load path, from the vessel through to the interface with the civil structure (see Fig 7) were estimated by NNC for a variety of load conditions in accordance with the requirements of the SFS for the Shiplift mechanical design. These included:
- normal operating conditions
- fault conditions involving the static and dynamic failure of the main transverse beams of the platform
- loading conditions imposed by a seismic event.

41. The estimated enveloping loads contained an allowance for the possible increase in load due to docking and vessel dimensional tolerances.

42. *Structural integrity analysis* The structural integrity of the major platform components under the estimated enveloping loads was substantiated in this phase.

43. The analysis and assessment methods adopted were the most appropriate for the individual component and loading

condition under consideration. The analyses ranged from simple hand calculations to complex linear and non-linear numerical methods and included fatigue and fracture assessments. The detailed nature of the finite element models used to analyse complex components such as the articulated and rigid bays is illustrated by Fig 8 which shows the ABAQUS (Ref 1) finite element representation of the rigid bay used by NNC for this analysis.

44. <u>Global response analysis</u> The global loading on the major platform components when supporting pre-defined vessel types under the various loading conditions required by the safety case was calculated in this phase.

45. A series of static and dynamic, linear and non-linear, finite element analyses was carried out for this purpose using a global model of the Shiplift platform and vessel. This global model was created using the ABAQUS sub-structure technique which offered a cost-effective state-of-the-art technique for capturing the detailed response of individual components in a global analysis (Ref.2). This method condenses detailed substructures into superelements which are subsequently used like building blocks to construct the global model. Where possible, the superelements for use in the global Shiplift model were developed from the detailed finite element models used in the integrity analysis to ensure consistency between the two models.

46. <u>Load reconciliation exercise</u> The loading information obtained from the global response analyses was reconciled with the estimated enveloping loads used in the structural integrity analysis in this phase. This was achieved by:
- A direct comparison between the global analysis loads and the enveloping loads used in the structural integrity analysis
- An assessment of margins and conservatism where the global analysis loads exceeded the structural integrity load assumptions.

47. <u>Implementation of the strategy</u> The structural analysis strategy was implemented successfully by NNC. There was some iteration between the various phases as the results from the global analyses became available, in particular when the results from the global analyses demonstrated that the estimated enveloping loads were unnecessarily conservative. An example of this was the allowance included initially in the estimated enveloping loads to account for docking and dimensional tolerances, which had been themselves revised during the development of the structural integrity safety case.

48. The strategy was implemented in parallel with the design of the major platform components. Minor design modifications were required to satisfy the safety case but these were readily incorporated before the design was finalised, with a minimum of programme delay and cost implications.

49. The nature and extent of the analytical work performed by NNC to support the safety case for the Shiplift platform are best illustrated by describing the analytical work that was carried out to support the safety case for the Shiplift platform control and protection systems.

50. <u>Analysis in support of the shiplift control and protection systems</u> The Shiplift control and protection systems are intended to safeguard the moving platform following a fault or hazard. The analytical work carried out to substantiate the performance of these systems, following a fault which affects

the hoists during raising or lowering operations with the Vanguard vessel on the platform, is described below.

51. **Safety Case Requirement** The control and protection systems will safeguard the platform following a hoist fault as follows:
- The load monitoring system continuously monitors the hoist loads
- The increase in hoist loads following a fault, as load redistribution takes place, will be monitored by the load monitoring system
- The protection system will remove the current from the hoist motor and apply the brakes when any one of the hoist loads reach a pre-defined level, namely the 'trip load'
- The platform is arrested in a safe and stable condition before critical platform loads are exceeded.

52. The objective of the structural analysis was to substantiate the above protection process by examining the platform loading in response to the fault, recognising the performance characteristics of the protection system, i.e. the time required by the protection system to apply the brakes and arrest the platform following a 'trip'.

53. **Outline of structural analysis** A series of global response analyses was undertaken to derive the rate of increase in platform loading following the hoist faults. Forty analyses were required to bound the possible hoist fault conditions which required safeguarding by the control and protection systems. These faults ranged from the failure of a single hoist to the loss of a power supply causing the failure of 16 of the 96 hoists. The large number of analyses was necessary because of the uneven spacing of the cradles over the platform.

54. The global finite element model of the Shiplift platform and vessel was used in the analyses. Minor modifications were made to the rigid bay superelement containing the 'failed' hoists in order to permit the required hoist failures to be modelled. Both linear and non-linear analyses were carried out. The latter were required to model the possible loss of contact between the cradle and vessel or between the cradle and rail.

55. The hoist failures were simulated by restraining the 'failed' hoists and lowering or raising the rest of the platform as appropriate.

56. The analyses produced information on the platform loading before and after the fault, including the rate of increase in the loading.

57. **Derivation of trip load settings** This platform loading information was used to derive trip load settings in the relatively narrow window that would avoid overloading the major platform component in the load path after a fault, whilst allowing platform operation in normal circumstances.

IMPLEMENTATION

58. The safety case framework and the development of the detailed requirements for the structures and equipment were successfully established by NNC by 1987. This stage required the production of 7 Design Concept Documents and 31 System Functional Specifications to cover all the new facilities. It was important to achieve this stage early in the project to ensure that the supporting Technical Reports were pertinent and complete. The structured approach ensured that requirements on

SAFETY CASE DEVELOPMENT

the plant were neither overstated nor understated. Some relaxations to MOD's specifications were negotiated as a result.

59. Thereafter the safety case development was focused on the production of the Technical Reports mainly by the Consultants to synopses prepared by NNC. The timescale for this stage was largely determined by two factors. The first factor was the construction, procurement and installation programme, since it is important that the safety case reflects the as-built design. The second factor was the extensive review process adopted by MOD Safety Authorities. In all some 130 Technical Reports were required, with each taken through several updates.

60. Potential difficulties that were overcome by NNC's experience and structured approach included the backfitting of the safety cases to pre-existing designs, and the necessary involvement of the large number of design consultants and contractors associated with the projects. In some cases it was necessary for NNC to provide closely detailed guidance, but this has ultimately been to the benefit of the organisations involved.

REFERENCES

1. Hibbitt Karlson and Sorensen Inc., ABAQUS User Manual, Version 4.9.

2. Curley, J. A., 'The use of superelements for seismic analysis', Proc. 7th UK ABAQUS User Group Conference, Cambridge, September 17th & 18th, 146-153.

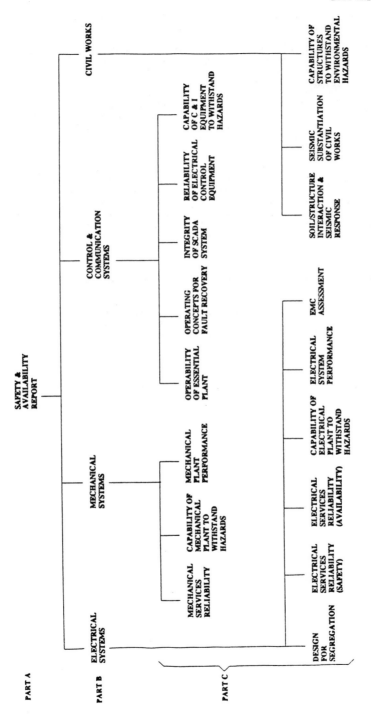

CSB FASLANE: NORTHERN UTILITIES BUILDING SAFETY CASE   FIG. 1

SAFETY CASE DEVELOPMENT

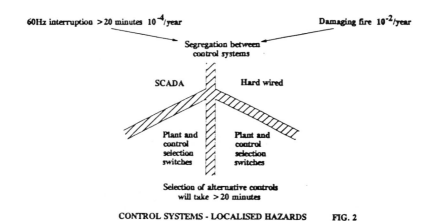

CONTROL SYSTEMS - LOCALISED HAZARDS   FIG. 2

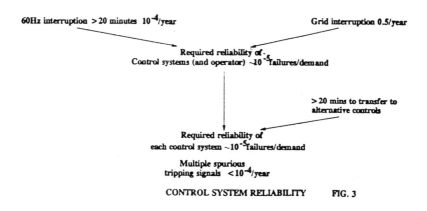

CONTROL SYSTEM RELIABILITY   FIG. 3

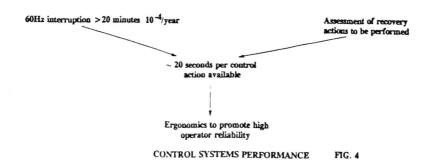

CONTROL SYSTEMS PERFORMANCE   FIG. 4

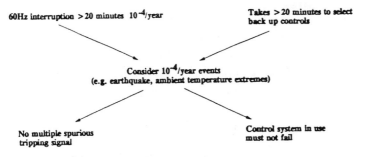

CONTROL SYSTEMS - WIDE-SPREAD HAZARDS   FIG. 5

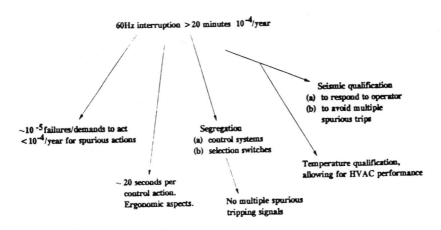

CONTROL SYSTEMS - SCADA SUMMARY OF SAFETY REQUIREMENTS   FIG. 6

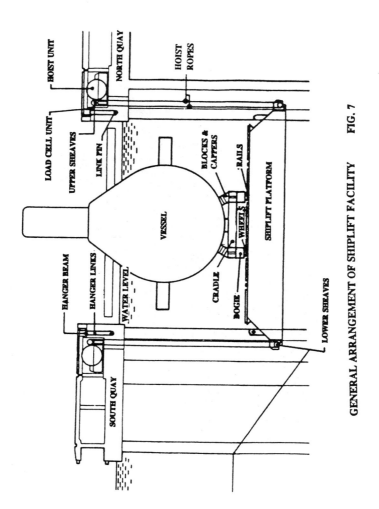

FIG. 7 GENERAL ARRANGEMENT OF SHIPLIFT FACILITY

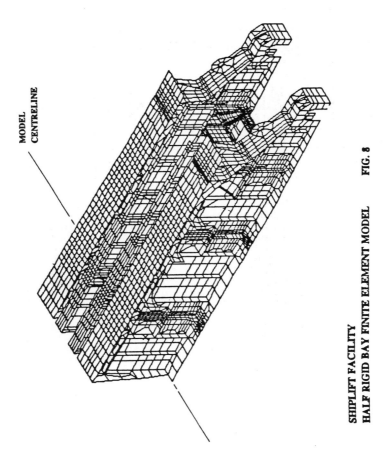

SHIPLIFT FACILITY
HALF RIGID BAY FINITE ELEMENT MODEL   FIG. 8

The Faslane reclamation

J. H. WOTTON, Managing Partner, Crouch Hogg Waterman, and
D. K. BELSHAM, Area Manager, Edmund Nuttall Ltd

SYNOPSIS. This paper discusses the design and construction of the principle bulkhead and the related structures forming the seaward margin to the major reclamation within the existing Southern Development Area at Faslane. The seismic design condition and the innovative solutions applied to ensure compliance with nuclear safety case requirements are discussed. The construction methods deployed to meet and assure the compliance with the specified requirements are described.

INTRODUCTION
1. The north and south lagoons located within the Clyde Submarine Base, Faslane, were infilled as part of the Southern Area development. Figure 1 highlights the proximity of the two lagoons to operational berths 2 to 5.

2. A comprehensive site investigation within the lagoons was undertaken early on in the design programme which revealed chlorite schist bedrock overlain by up to 6 metres of soft alluvial silt with a thin layer up to 3 metres of stiff glacial till present in the north lagoon only.

3. Feasibility studies were undertaken to determine the most appropriate form of construction for the main bulkhead based on the initial non-seismic design requirements. The depth of water and proximity to the existing jetties precluded a rock slope whilst the lack of overburden ruled out a conventional anchored bulkhead.
Cellular cofferdams of 12.088 and 15.240 metre diameter were chosen for the main bulkhead with anchored sheet piled walls forming the central and end closures. Figure 1 illustrates the location of each of the elements of the main bulkhead and their relationship to the various building structures which were subsequently constructed on the reclamation.

CELLULAR COFFERDAMS DESIGN

4. The cellular cofferdams are located immediately behind the existing operational jetties. In addition to acting as the main seaward retaining bulkhead to the reclamation, they also provide ground support to the main service distribution tunnel, electrical substations and NTD support buildings comprising the jetty modernisation programme in the Southern Area of the Base. These structures were supported on piled foundations driven into and through the cells and adjacent reclamation. Figure 2 illustrates the configuration.

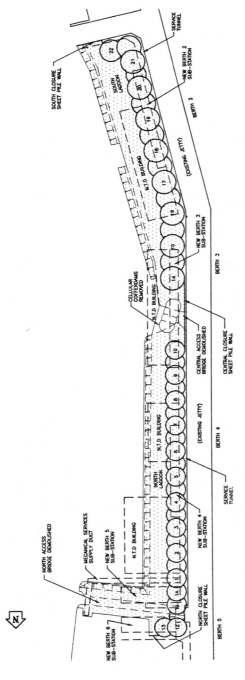

Fig 1. Location Plan

FASLANE RECLAMATION

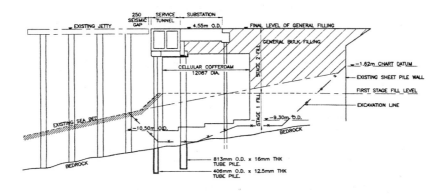

North Lagoon

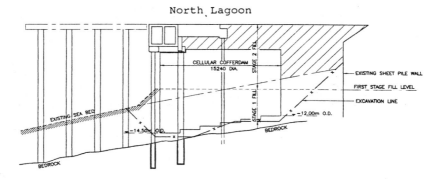

South Lagoon

Fig 2.   Typical Cross Sections

5.   At the commencement of design there was no requirement for any of the structures in the Southern Area to have a seismic withstand capability. Accordingly, the cellular cofferdams were analysed conventionally to determine the limiting factors of safety under a number of accepted failure modes, i.e., sliding, overturning, interlock failure, foundation failure, internal shear, tilt and slip circle.
Analysis was carried out using conventional earth pressure theories with due account being taken of the various applied loadings acting on the cells. The results of this work generated the factors of safety stated in Table 1.

6.   Subsequent to contract award, the requirement for aseismic capability of all nuclear safety structures in the Southern Area was confirmed and a programme of redesign work implemented.

7.   For the cellular cofferdams the acceptance criterion was defined as a minimum factor of safety of 1.10 for any failure mode.

Table 1. Non Seismic Factors of Safety

Failure Mode	North Lagoon 12.088m. cell Factor of Safety	South Lagoon 15.240m. cell Factor of Safety
Sliding	2.94	2.61
Overturning	5.25	6.07
Interlock failure	5.14	3.29
Foundation bearing failure	5.70	5.81
Internal Shear	6.05	8.00
Tilt	6.72	7.30
Slip circle	2.14	Not Applicable

8. The reference earthquake was defined by zero period ground response accelerations of 0.2g and ± 0.135g in the horizontal and vertical directions respectively.

Equivalent seismic static earth pressure forces acting on the wall were derived for these accelerations using the Mononobe and Okabe equations as modified by Matsuzawa et al (Ref. 1).

Equivalent seismic static forces for all other masses contained within the seismic soil failure wedge were calculated, in the case of structures supported on piles in the horizontal direction only whilst other structures or surcharge loads supported directly on the ground surface in both the horizontal and vertical direction.

The limiting design condition conservatively assumed the acceleration of all contributing masses within the failure zone as acting in phase.

The various forces acting on the cell are illustrated in Figure 3.

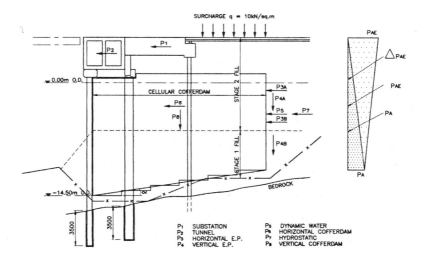

Fig 3.   Cofferdams - Load Diagram

9. Preliminary work suggested that because of the effects of seismic uplift and the consequent loss of weight of the cell and hence reduction in base friction together with the natural dip of the bed rock across the cell sliding failure would be critical with the factor of safety close to or less than unity. Accordingly, a very careful reassessment of the factors affecting cell stability was carried out. In particular, the following were reviewed in detail:-

- shear strength of foundation strata
- density and shear strength of backfill materials
- wall friction

10. The original design concept had foreseen the need to avoid the risk of distortion during filling and the consequent problems of clutching the connecting arcs by specifying the removal of soft silts and the excavation of a stepped formation in the relict glacial tills. Granular fill was then to be placed to a level platform from which cells were then constructed and backfilled.

11. Critical reassessment of the site investigation results suggested that, given sufficient sea bed excavation, the cells could be founded in strata having the minimum limiting strength values set out in Table 2.

Table 2. Foundation Strata Strength Parameters

Location	Founding Stratum	Angle of Friction ϕ	Unconfined Compression Strength C_u
North Lagoon	Very stiff boulder clay	-	150
N & S Lagoons	Very dense gravel	43	-
S Lagoon	Bedrock	45	-

The figures represent typical ultimate values. For the purpose of calculation for cells founded on either very dense gravel or bedrock a value of 40° for internal friction was adopted, corresponding to a coefficient of friction on the base of 0.84 i.e., tan 40°. For cells founded in the very stiff boulder clay, 150kN/m² was taken for the undrained cohesive strength.

12. The backfill and infill materials to the cells were reviewed with respect to density, angle of internal friction and angle of wall friction and it was concluded that the original specification should be modified to incorporate a requirement which defined the angle of internal friction as not less that 40°.

The specification defined an acceptance criteria of at least 37½° for the natural angle of repose of materials in loosely formed stock piles being equivalent to an angle of internal friction of at least 42½° for moderately compacted material at moderate confining pressures such as the insitu condition.

13. In order to enhance the value of wall friction acting on the wall and therefore to increase the vertical component of earth pressure positive measures were taken to roughen

artificially the earth retaining face of the straight web piling. This took the form of horizontal bars welded at 150mm. vertical centres to all piles on the earth retaining face, thus promoting shear failure through the backfill and not at the soil-wall interface.

A value of wall friction equal to the internal angle of friction of the fill, i.e., 40° was therefore adopted.

14. The results of the seismic analysis generated the factors of safety presented in Table 3.

Table 3. Seismic Factors of Safety

Failure Mode	North Lagoon Factors of Safety	South Lagoon Factors of Safety
Sliding	1.06	1.04
Overturning	1.43	1.53
Interlock failure	1.62	1.15
Foundation failure	1.80	2.03
Internal shear	1.18	1.25
Tilt	1.16	1.25
Slip circle	1.46	Not Applicable

With the exception of sliding, all factors of safety were found to be greater than the 1.10 requirement.

15. In order to enhance the resistance to sliding, the inner line of service tunnel support piles installed within the cell/arc envelope, originally driven H piles, was modified to 813mm large diameter tubular piles drilled and socketed into bed rock.

16. The socketed tubular piles behave as deeply embedded passive anchors and contribute significantly to the overall factor of safety against sliding increasing the factor of safety to greater than 1.10.

The theoretical passive pile reaction required to provide a factor of safety against sliding of 1.10 is approximately 700kN corresponding to a displacement of about 30mm. The setting out of the cells and topside structures took full account of these anticipated displacements.

CENTRAL CLOSURE WALL DESIGN

17. The central closure wall is located between the north and south lagoons in way of the existing central access bridge to Berth 3. Figure 4 illustrates the situation.

18. Design of the new wall was complicated by the presence of the access bridge and in particular its landward abutment which was of cellular cofferdam construction built in 1964. It was recognised that the proximity of this structure to the new wall would influence the form and location of the seismic failure wedge which would inevitably propagate from the heel of the old cofferdam rather than at the new wall with the consequent increase in horizontal load onto the new structure. Additionally, the absence of records for the abutment structure cast doubts on its ability to withstand the reference earthquake loading. It was therefore decided that this structure should be removed prior to constructing the new wall.

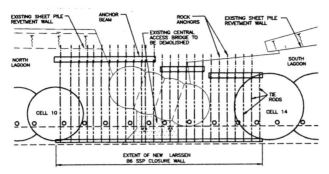

Fig 4. Central Closure Wall Layout

19. Derivation of the various seismic forces acting on the wall has already been described above. However, the new sheet pile wall also provides the vertical support to the new service tunnel. Accordingly, the wall was designed to carry simultaneously all applied horizontal in-phase equivalent static loadings in bending together with vertical compression loading from the service tunnel support. Figure 5 illustrates the various forces acting on the structure, whilst Figure 6 illustrates the design cross section.

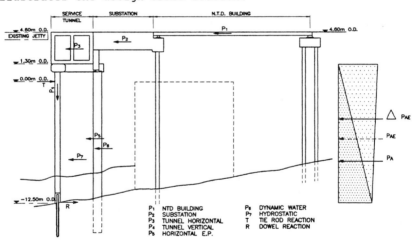

Fig 5. Central Closure Load Diagram

20. The sheet piled wall is conventionally designed as a simply supported beam spanning vertically between the tie rod and rock head where the piles are restrained by toe dowels. Passive pressure effects were discounted in the design of the piles and toe restraint system.
Larssen no. 6 box piles 22.1mm thick in grade 43A steel were adopted for the main bulkhead, painted with 300 microns of isosyanate coal tar epoxy paint.

Under the worst combination of bending and axial compression, the piles are stressed to 68% of yield.

At the end of the 50 year design life stress levels rise to 86% of yield based on reduced theoretical section properties derived from metal loss data published by British Steel (Ref. 2).

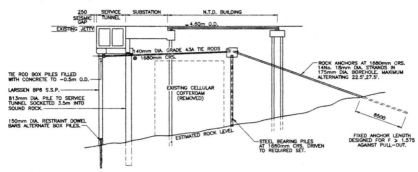

Fig 6.   Central Closure Section

21.   The pile toes are restrained using a system of dowel pins installed in alternate box piles.

In designing the reaction system, the dowels are assumed to transfer load from pile to bed rock in a combination of bending and shear.

Sensitivity studies were carried out to consider variations in the location of plastic hinging in the dowel pin which resulted in 150mm diameter grade 50B pins being adopted.

22.   A conventional passive anchor wall was considered but rejected because of difficulties associated with its seismic qualification.

The anchor system was therefore based on the use of inclined rock anchors. However, because of the risk of damage to deeply buried rock anchors from piling for future topside structures, the rock anchors had to be located clear of the foot print of these future buildings.

Figure 5 illustrates the anchor system which comprises 140mm diameter grade 43A tie rods at 1.680 metre centres spanning between the new wall and a reinforced concrete ground anchor beam. The ground beam is supported on steel H piles driven to refusal on rock and forms the reaction point for the tie rods and the inclined rock anchors.

The rock anchors were installed at 1.68 metre centres offset from the tie rods by half a pitch and at alternate inclinations of 22½° and 27½° to avoid the possibility of a continuous pull out wedge failure in the underlying rock. Anchors were made up of 14 No. 18mm diameter 7 wire drawn strands, giving a factor of safety of 3 on the minimum unfactored seismic load. The fixed anchor length was calculated to be 6.5 metres.

All anchors were satisfactorily tested to 1.575 times the maximum seismic load.

23.   Provision was made for differential displacements between the closure wall and the adjacent cells by the construction of end closure boxes and the incorporation of full articulation joints in the service tunnel.

FASLANE RECLAMATION

SOUTH CLOSURE WALL DESIGN

24. The location of the south closure wall is indicated on Figure 1.

25. The design of this wall is similar in concept to that already described for the central closure. There are however some differences in detail which result from differences in geometry and applied loading.

26. A Larssen 32W alternate sheet and box pile wall in grade 50A material was adopted. Toe restraint is provided by 150mm diameter grade 50A dowel pins to every box pile.
The anchor system utilises 100mm diameter grade 50A tie rods at 2.10 metre centres connected via a reinforced concrete piled ground beam to inclined rock anchors also at 2.10 metre centres offset half a pitch. The rock anchors comprise 16 No. 15mm diameter 7 wire drawn strands and have a fixed anchor length of 4 metres.

CELL CONSTRUCTION

27. Prior to construction of the circular cell cofferdam, the specification called for preparation of the existing sea bed as described in paragraph 10 above. The effect of the underwater excavation upon adjacent structures had to be considered with temporary support works designed and constructed where required.
This work was made more complex by the imposed requirement for the founding strata to have an insitu strength equivalent to an angle of internal friction of 40° or a cohesive strength of 150 kN/m². In addition the work was affected by the presence of uncharted buried timber pile stobs which needed to be removed.

28. Temporary works were required to support the adjacent existing sheet pile wall shown on Figure 7. Details of this wall were unknown, and accordingly a scheme was implemented depending upon the anticipated excavation profile. In the area of cell Nos. 6 to 10 it was predicted that the existing wall would still retain material but required support. This was provided by H piles driven in front of each alternate pile span position and penetrating 1m into bedrock. The heads of the H piles were tied back by rock anchors, details of this scheme being shown in Figure 7.

29. Following construction of the temporary support works, the cell bed area was initially excavated north to south using a combination of shore based and floating plant. All excavated material was transferred into a bottom dump barge and then deposited at sea in a licensed disposal area.
Once the pile stobs had been initially located, a team of divers airlifted around the timber stobs for a depth of approximately 1.5m or until sound timber had been uncovered. The airlifting was carried out from within a box to protect the divers from ground collapse around the pile. After several trial methods of removing these stobs, including direct line pull and uplift floatation bags, a hydraulic jawed device welded to an ICE 216 vibrator was devised. Known as the "green gizmo" this proved to be an efficient system, capable of extracting a 3 metre long timber pile in under fifteen minutes.

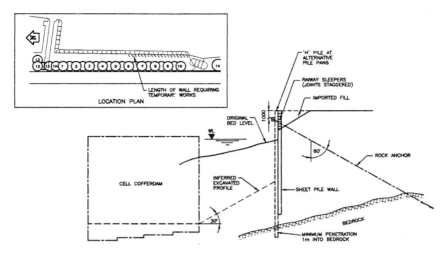

Fig 7. North Lagoon Temporary Retaining Wall

After removal of the pile stobs, the cranes were changed back onto grabbing duties and completed the excavation to the founding stratum. A pontoon mounted shell and auger site investigation rig carried out tests on the exposed surface of the final excavated level to prove compliance with the insitu strength requirements. Bed surveys were then carried out to ensure that the formation profile was acceptable.

30. The stage one fill (60mm down) was deposited via a crane operated rock skip into the required area of the cell bed. Check surveys were carried out during the final stages of the filling operation to direct the fill into specific areas. Use of a 2.5m x 2.5m x 0.5m concrete slab lowered to predetermined levels and then swept over the cell bed area provided confidence that the final level of the fill material had been placed to the specified 0.5m tolerance.

31. The circular cofferdams shown in Figure 1 were used to reclaim the lagoons. These comprised 11 No. 12.088m diameter cells together with arcs and 2 No. ¾ cells in the north lagoon, and 8 No. 15.240m diameter cells together with arcs and 2 No. ¾ cells in the south lagoon. Frodingham SW1A piles were specified varying in length up to 16.4m long.

32. An internal template weighing 12.1 tonnes was designed to dimensions of 12.088m diameter by 3m deep. It was constructed in four quadrants that were assembled into a single unit on land and then lifted onto a specially adapted uniflote pontoon. This was then pulled into position and maintained on station by the use of four winches. The spud piles were pitched and driven through the granular fill onto rockhead using an ICE 233 vibrator. Positional checks were carried out throughout. Hoisting caps were then inserted into the top of the spud piles and 4 No. tirfors used to lift the template clear of the pontoon. After removal of the pontoon, the template was lowered until the upper surface was at mid tide level. The four Tee piles forming the junction between the cell and its adjacent arcs had been adapted such that a section

of SW1A pile clutch had been welded onto its inside face. These junction piles were then pitched into matching "zipper" connections on the template. A lightweight outer frame was then lowered between the junction piles and set to the same upper level as the inner template. The circular cell sequence of construction is shown in Figure 8.

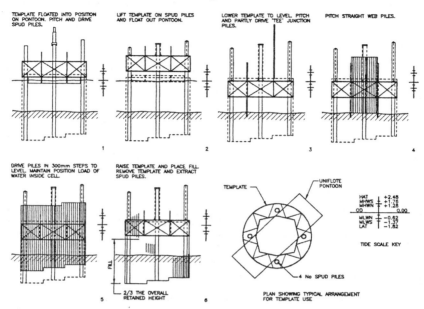

Fig 8.   Cell Construction Sequence

33. Positioning of the ¾ cell template involved location, at high tide, of the four quadrant configuration template above and overlapping the adjacent cell. The four spud piles were then driven, one of them through the adjacent cell fill. After removal of the pontoon, the overlapping quadrant was removed and the remaining ¾ template lowered to the level suitable for pile pitching.

34. The 12.088m diameter template was designed such that, through bolting on additional members, it then served for the 15.240m diameter cell construction. Its weight in this configuration was 15.4 tonnes. The cells were constructed using a 1405D crawler crane working either from land, from access jetties or from adjacent completed cells working end over end. The straight web piles were lifted from the horizontal with two lifting ropes to avoid excessive bending and pile damage. These were then pitched around the template working from the T junction piles until each quadrant was fully pitched and "shaken down" such that the whole quadrant of piles were sitting on the cell bed preparation fill.

35. Working around the cell, driving each quadrant in turn, the piles were advanced in pairs into the stage one fill in 300mm increments, by use of an ICE 223 vibrating hammer. At each of the junction piles a pair of straight web piles, external to the cells, were pitched and driven into the fill

prior to cell filling. Piles were driven to predetermined levels.

36. As soon as practicable after completion of pile driving, the template was hoisted up approximately 1 metre using the tirfors. Care was taken to avoid a negative head of water inside the circular cell in order to maintain stability before filling. Stage two fill material was placed into the centre of each cell by the 1405D crane using a 12 ton rock skip filled by a loading shovel. Fill was placed to a level 0.5m from the underside of the template which was then removed onto its uniflote pontoon. Infilling then continued until the cell was full. Care was always taken to place the fill centrally into the cell and to check the level of fill around the cell perimeter at regular intervals to avoid concentric loading. Spud piles were then extracted and straight web piles trimmed to level.

37. For the purpose of support to the service tunnel and to provide deeply embedded passive anchors to the circular cells, bents of 406 OD x 12.5 thick and 813OD x 16 thick tubular piles were specified at 5.4m centres. These piles were required to be inserted and concreted into a 3.5m deep socket into the bedrock.

The piles were installed using temporary casings of 660mm diameter and 1067mm diameter allowing socket diameters of 600mm and 1000mm. These dimensions provided for the 20mm external rings specified for the base of each pile and appropriate concrete cover.

38. A hydraulic piling gate was designed and fabricated that allowed the pitching and driving of one bent of adjacent 406mm and 813mm diameter piles. Using a land based 1405D crawler crane seated on the completed circular cells, the casings were pitched in front of the cell and driven through the Stage 1 fill in the case of the 406 diameter piles casing and through the cellular cofferdam fill for the 813m diameter piles casing. Use of a BSP 357 hammer in 5 tonne mode allowed the casing to reach rockhead.

39. Following removal of the hydraulic gate and subsequent temporary bracing of the 406mm diameter casing, the material inside the casing was removed using a Wirth B5 reverse circulation pile top drilling rig. The same equipment was then used to drill out the rock socket to a depth of 3.5m. Time taken to drill the socket varied dramatically depending upon the presence or otherwise of quartz bands in the schist bedrock.

After removal of the B5 drill, concrete was tremied into the rock socket and the pile placed within the casing. In order to maintain the pile in its correct location, the casing was surveyed and timber packers fixed to the pile such that when it was positioned within the casing its location met the specification requirements. Using an internal drop hammer inside the casing the final pile was then driven to the bottom of the socket through the wet tremied concrete. The casing was then extracted twenty four hours after pile driving and bracing fixed to the 406mm diameter pile. Construction of the reinforced concrete service tunnel was then able to proceed.

FASLANE RECLAMATION

CENTRAL CLOSURE WALL CONSTRUCTION

40. The permanent and temporary works in this area are shown in Figures 4 and 9 and comprised demolition of the existing central access bridge and extraction of the existing circular cells, sea bed excavation and fill, installation of a Larssen BP 6 box pile wall between cells 10 and 14 with toe dowels, tie rods from the box pile wall to piled anchor beams and rock anchors individually test loaded to 2290 kN. These operations were required to be carried out in a confined area, to as fast a timescale as possible, and maintaining access to the existing jetty for the MOD (Navy).

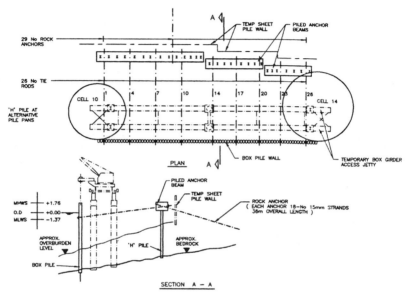

Fig 9. Central Closure Construction Sequence

41. Demolition of the existing central access bridge was carried out by placing temporary supports, stitch drilling and then removing large sections by use of heavy craneage. Supporting piles were then extracted.

Using a crane grab the sea bed between cells 10 and 14 was excavated to hard formation levels equivalent to that experienced in the adjacent cell bed preparation. Granular fill was then placed up to the blinding level of the anchor beam.

42. The contract requirements called for the box pile wall to have toe fixity provided by 150mm diameter dowels 4m long. In order that no bending was transferred to the dowels it was essential that the box piles were driven fully to sound rock. A combination of potentially hard driving due to clutch friction and any obstructions or boulders in the overburden meant that the requirement to ensure the pile toes were at rockhead would be difficult to guarantee. For this reason, a predrilled trench formed by overlapping sockets drilled by a Wirth B5 and backfilled with sand was constructed in advance along the line of the box pile wall. In order to carry out

this operation and provide access for the associated equipment and maintain concurrent work on the anchor beam, a temporary access jetty was designed and constructed. This comprised 4 No. 25m long box girders each spanning 21m. The centre supports were 2 No. 1067 diameter tubular piles driven to refusal. The outer supports were 4 No. UBP again driven to refusal.

43. A mobile rail mounted electrically driven carrier was then constructed on the access jetty. This was fitted with fixed drilling and piling platforms. A 1405D crawler crane then drove temporary 1067mm diameter casings through the overburden and a Wirth B5 rig drilled 500mm into sound rock at each casing in turn. After drilling, the casing was filled with sand up to 1m above sea bed level and then extracted. An American Hoist 5299 crawler crane with fixed leaders then drove the box piles wall into the sand filled trench using a Delmag D30 diesel hammer. The sand was airlifted from inside the box piles and a second carrier supported a drilling rig to drill a 250 diameter hole 2.5m into rock below the toe of the pile base. The 150mm diameter steel dowel was lowered into the hole and grouted up to give 150mm cover to the head of the dowel.

44. Simultaneously, the bearing piles supporting the anchor beam were driven to refusal together with a temporary sheet pile wall on their landward side that would subsequently be used to spread the anchor test load.
Trial rock anchors were constructed and tested against the sheet pile wall, following which the permanent anchors were installed off a purpose designed skid rig to meet the $22\frac{1}{2}°/27\frac{1}{2}°$ parameters. Each anchor was test loaded to 2290 kN. Following anchor installation, the anchor beam was cast on top of the UBP support piles and around the rock anchors. After sufficient curing of the beam, each anchor was locked off at 150 kN. Prior to installation of the 140mm diameter tie rods the existing circular cells were extracted. The tie rods were then hung from the access jetty beams and tightened. Strain gauges were attached and monitored during backfilling operations. A 500mm diameter split duct was placed around each tie rod to ensure that any settlement of the fill would not induce bending into the tie rods. Following final backfilling to cell Nos. 10 and 14, service tunnel construction was able to proceed.

ACKNOWLEDGEMENTS

45. The assistance and co-operation of the Ministry of Defence in the preparation of this paper is gratefully acknowledged.

REFERENCES

1. MATSUZAWA et al; Dynamic Soil and Water Pressures of Submerged Soils; Journal of Geotechnical Engineering, ASCE, Vol. III No. 10 October 1985.

2. BRITISH STEEL CORPORATION; Piling Handbook; 1988.

Design and construction of the Shiplift and Finger Jetty

W. J. PATERSON, Director, R. DUNSIRE, Technical Director, Babtie Shaw and Morton, and D. A. BLACKBURN, Project Director, Trafalgar House Construction

SYNOPSIS. This Paper describes the principal civil engineering design and construction aspects of the Shiplift and Finger Jetties. These two marine jetties provide 3 conventional berths and the Shiplifting facility within the Northern Development Area of the Clyde Submarine Base, Faslane. Nuclear Safety criteria dictated that the structures were designed to resist extreme hazard loads including earthquake loading. Construction of the jetties was technically challenging, particularly in pile installation and jetty deck construction, and was made more onerous by tight programme restraints.

INTRODUCTION
1. Located in the Faslane Northern Development Area, the Shiplift package provides the primary support facility for the Trident submarine fleet. The package comprises a Finger Jetty and Shiplift, together with a utilities building (the NUB) providing power generation and support services, and infrastructure works linking these to the jetties. Three fully serviced deep water berths are provided, one either side of the Finger Jetty and one alongside the Shiplift Jetty. These Berths feature portal cranes running on the jetties. The centre piece of the package is the Shiplift itself which provides a covered, and fully serviced, maintenance facility, with two overhead cranes. Submarines are lifted into the facility by means of the shiplift platform which spans between the quays of the jetty. The Shiplift package included provision of all M&E services and equipment necessary to support the facility.
2. Babtie Shaw & Morton were responsible for the Civil and Structural design of the two jetties and the NUB and were also the Lead Consultants for the Shiplift Package heading a multi-discipline design team consisting of six organisations.
3. Trafalgar House Construction (Major Projects) was the main contractor for the whole of the works including the M&E installation. The latter included an element of design work.

Fig. 1 - Shiplift Structure Nearing Completion

DESCRIPTION OF SITE

4. The Northern Development Area (NDA) is the area immediately to the North of the established Clyde Submarine Base at Faslane. The NDA was previously the site of a ship breaking yard which had left parts of the site contaminated with asbestos. A previous contract had made safe contaminated material by collection and storage on site.

5. A site investigation of the NDA was conducted by Wimpey Laboratories Ltd. prior to commencement of the design. On the sites of the Shiplift and Finger Jetties 24 and 8 marine boreholes were sunk respectively. A seismic profiling exercise and laboratory testing were also carried out.

6. Seabed levels ranged from around -6m AOD at the existing sea wall to -25m AOD with rockhead between -18m AOD close to shore to -35m at the extremity of the structures. The rock was Phyllite, a fine grained schist, with quartz veins, with overlying material consisting of a layer of soft silty clays up to 10m thick over thinner layers of sands, gravels, and boulder clay. Very stiff boulder clays predominated at the Finger Jetty where this layer was up to 16m thick.

DESCRIPTION OF STRUCTURES

7. The substructures of the Shiplift and Finger Jetties feature similar forms of construction. Cast insitu reinforced concrete jetty decks are supported by steel tubular piles which are either driven to rockhead or socketted into rock. The Shiplift is the much larger structure measuring 285m in length and 120m in breadth at its widest point while the Finger Jetty measures 223m by 24m. The piling system supporting the Shiplift jetty consists of some 820 steel tubular piles varying in diameter from 600mm to 1000mm. The pile lengths vary from 17 metres at the landward end of the structure to 55 metres at the seaward limit.

The 214 piles on the Finger Jetty possess a similar range in diameters and also progressively increase in length from the shore to a maximum of 54m. On both jetties the majority of piles are vertical, with some pitched at a rake to provide horizontal stability.

8. The raking piles are arranged in pairs so that loads are in balance between each set of twin rakers. On both jetties there are approximately twice as many transverse raking piles as longitudinal rakers.

9. The reinforced concrete decks of both jetties are cast insitu and are of a monolithic construction. Generally the decks are constructed in two levels, the top level providing vehicle and pedestrian access onto the jetties while the lower level provides duct space for services. Integral with the decks are a series of walls, so the jetties are in general, of a cellular form. For construction of the jetty decks, Grade 40 concrete has been used throughout. A PFA mix was chosen for the concrete to aid workability and to provide a dense, durable deck, characteristics which are essential in a marine environment. On concrete surfaces exposed directly to the marine environment the cover to reinforcement is 75mm. There are some 26000m^3 of concrete in the Shiplift jetty decks and 7600m^3 in those of the Finger Jetty.

10. On the Shiplift jetty there are a number of distinct areas of the decks which fulfil different functions. These areas are shown in Figure 2.

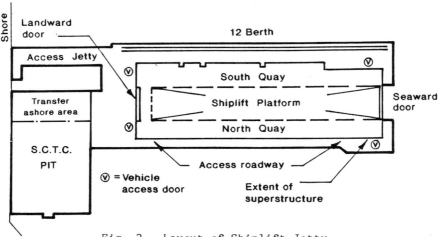

Fig. 2 - Layout of Shiplift Jetty

11. The Access Jetty, Approach Apron, South and North Quays provide services and access to the Shiplift platform and to any vessel on the platform. The North and South Quay winch chambers each hold 46 No. 400T winches which support the main section of the Shiplift platform and a further 4

winches supporting a road bridge at the seaward end.
The Side Cradle Transfer Carriage (SCTC) pit is a lower area of the structure which houses a shuttle carriage. Its function is to transfer wheeled cradles used for docking a vessel from a marshalling yard onshore to the Shiplift platform.

12. On the seaward half of the jetty the deck separates into the South and North quays in a tuning fork type configuration. The area between these quays is occupied by the Shiplift platform itself which was designed by NEI Syncrolift. The platform is capable of lifting a vessel for maintenance (refer to Fig. 3).

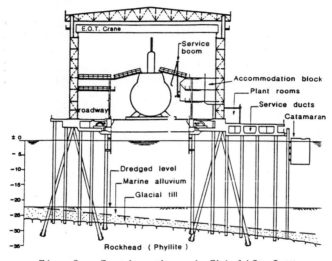

Fig. 3 - Section through Shiplift Jetty

13. A steel clad superstructure spans between the South and North quays and covers the Shiplift platform area of the jetty. The Superstructure, measuring 185m by 51.5m in plan and is 36m in height to the portal knee, houses accommodation and plant rooms as well as providing high level access to any vessel docked on the Shiplift platform.

14. Latticed steel portals provide the principal structural elements within the superstructure. These portal frames provide lateral stability to the Shiplift cover while longitudinal stability is achieved using three reinforced concrete stair towers located within the building on its South side and corresponding structural steelwork braced bays on the North elevation. The column sections of the portals also support a crane rail for two travelling overhead cranes.

15. Both jetties feature berthing and mooring systems. Bridgestone cell fenders mounted to vertical fender piles provide the necessary berthing energy absorption.

SHIPLIFT AND FINGER JETTY

A series of bollards and capstans provide mooring points for the vessels. On the Shiplift, stainless steel rail-mounted lead-in trolleys provide mooring control of the vessels entering and departing the Shiplift basin.

Fig. 4 - Finger Jetty

16. In order to berth submarines to the jetties, three rectangular floating steel "catamarans" 'space off' the vessel from the jetty edge to ensure its hydroplanes do not come into contact with the jetty edge.

17. There are two separate buildings located on the centre line of the Finger Jetty deck. These run longitudinally and are steel framed with concrete stair towers acting as hard points within the structure.

18. Two railed portal cranes straddle the Finger Jetty buildings. The cranes run longitudinally and have a lifting capacity of 20 and 125 Tonnes. The largest crane imparts a dead load of some 1300 Tonnes on the jetty.

CIVIL ENGINEERING DESIGN

19. The jetty structures at Faslane are designed for normal operational and environmental loads but are also designed to cope with a number of extreme environmental and accident hazards.

Parameters for normal design loads

20. The design load cases for normal loads include the following:

a) Berthing and mooring loads
b) 1 in 100 year wind and waves
c) Crane pad loads from mobile cranes
d) Loads from portal cranes
e) Imposed loads from the Shiplift platform
f) Loads imparted on the Shiplift jetty during vessel transfer ashore

21. The above load types are common to both the Shiplift and the Finger Jetty with the exceptions being those unique to the Shiplift where the platform imposes high local loads along each quay edge (400 Tonnes per support pile). The landward part of the Shiplift jetty is also designed for transfer ashore loads. It is not presently possible to transfer a vessel ashore at Faslane since the necessary onshore developments have not been built but it may be a future requirement.

22. At the onset of design all requirements were assessed and open piled structures with cast insitu decks for both jetties were identified as being the most appropriate structural form particularly for the high integrity required for nuclear related use. The piles on the Finger Jetty are driven to rock while on the Shiplift all raking piles are socketted into rock as are a number of vertical piles. There is limited overburden on the site of the Shiplift therefore, unlike the Finger Jetty where skin friction from the overburden provides adequate resistance to raking piles in tension, the Shiplift rakers are socketed into rock to provide a tension anchor. The raking pile sockets are filled with concrete and vary in depth between 3.5 and 8.5 metres. The raking pile sockets feature a two stage under-ream which provides a mechanical key into the rock. External shear rings on the raking piles transfer loads into the socket.

23. The raking pile socket depths were dictated by the tensile loads in the raking pile and the depth of overburden at the site of the socket. Where there was little overburden, the tensile resistance of the pile was taken by the weight of the inverted rock cone which would be mobilised by the socket. Where significant overburden was in evidence the ballasting weight of the overburden was recognised in reducing the required depth of socket. The design load for tensile rakers results from safety case hazard events causing sway of the jetty. The Shiplift sockets have a typical ultimate resistance of 600 Tonnes. At the pile heads on both jetties, 40mm x 40mm square steel straps are welded to the outside of the pile wall and anchor the pile to the jetty deck.

24. The piles were designed to BS449 and later checked using BS5950. A coal tar epoxy paint coating protects the jetty piles against corrosion in addition to an impressed current cathodic protection system. A corrosion assessment study and maintenance procedures indicate the 40 year life of the piling system will be achieved.

25. Generally, thickened sections of the jetty decks span between piles or groups of piles and in turn, slabs span between these thickened areas. Most external areas of the jetties are designed for 100 Tonne pad loads from mobile cranes. For the Finger Jetty the portal cranes produce the most onerous bending moments in the principal deck elements.

26. The superstructure buildings are designed to withstand 1 in 100 year environmental loads and in the case of the Shiplift operating loads from the two 55 Tonne capacity electrical overhead travelling cranes. The superstructures are also designed to withstand hazard loads as part of the safety case assessment.

The Safety Case

27. As the jetty structures provide services to nuclear reactors of berthed submarines, it is necessary to demonstrate the inherent safety of the structures. Substantiation of the safety case included an assessment of the structures against extreme hazards which included

a) An earthquake acceleration of magnitude 0.2g horizontally
b) 1 in 10,000 year wind and waves
c) 1 in 10,000 year temperature effects
d) Explosions
e) Extreme tide and flood
f) Dropped loads from crane operations

28. For each of the hazard events the structural behaviour of the jetties had to be calculated. This often involved the use of complex analytical models of the structure to compute its dynamic response to the hazard loading. A design assessment was then conducted to establish the factors of safety on structural elements.

29. Factors of safety were used in the normal manner in designing the structures for operational loads and 1 in 100 year environmental loads. For safety case hazards, with recurrence periods of 1 in 10000 years the appropriate design targets were as follows. A partial load factor of safety of 1.05 was applied in conjunction with different material factors of safety. The relevant material factor of safety varied depending on the type of behaviour against which the element was being assessed. For ductile behaviour (e.g. a beam in bending) a material factor of safety of unity was chosen. For a brittle failure (e.g. punching shear through a slab) a material factor of safety of 1.3 was used while for a connection 1.5 was the apropriate material factor of safety. The philosophy then was to allow the structures to respond in a ductile manner to these extreme load conditions while ensuring that brittle mechanisms and connections had sufficient reserves of strength to resist the applied load.

30. For each of the hazard loads considered, the structural assessment was described in a report which was then reviewed by independent nuclear safety consultants.

31. It is not possible within the scope of this paper to provide information on all of the hazards events and the associated structural behaviour of the jetties.

What follows therefore are brief descriptions of some of the aspects associated with the seismic assessment of the structures.

32. The first stage in the seismic assessment was to analyse the dynamic response of the structures by the use of 3D finite element computer models. Structural analyses were carried out both at the design stage and after construction so that "as built" conditions could be assessed. The latter analyses used more detailed models which did identify local changes in dynamic response.

33. The design assessment of the structural elements to seismic loads was an extensive exercise. This was particularly so for the jetty piles where a number of factors had to be recognised. These included:

a) The influence of marine growth on the higher modes of oscillation of individual piles.
b) The actual pitch angle and orientation of installed raking piles.
c) Out of straightness tolerance on the pile.
d) The influence of jetty sway on the buckling capacity of piles.
e) Allowance for corrosion of piles over the design life of the structure.

34. An interesting example of safety case investigation is the effect of residual stresses on the jetty support piles.

35. In fabricated structural steel elements such as tubes, welds will contract on cooling inducing residual stress in the material. The effects of residual stresses on those jetty piles which were fabricated by spiral welding were uncertain and were therefore investigated. Information was available which demonstrated that tubes fabricated by welding together longitudinally seam welded cans possessed an acceptable residual stress pattern. It therefore had to be demonstrated that spirally welded tubes did not induce a worse residual stress. This was done by developing two computer models to investigate these effects. The first model represented a longitudinally seam welded tube. Residual stress patterns were simulated by inducing a contraction force along the element representing the weld inducing a tendon stress in the tube. A second model repeated the study but this time a hellical weld was represented in the finite element study. Residual stress patterns computed by the two models were compared and this demonstrated those from the spirally welded tube were less severe. These are illustrated in Fig. 5.

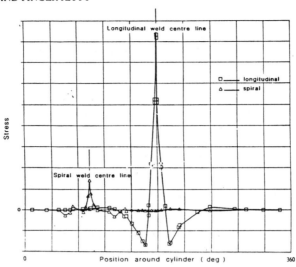

Fig. 5 - Residual Stress Patterns - Fabricated Steel Tubes
(Analysis conducted by NNC Ltd.)

PLANNING

36. Comprehensive and detailed planning of both design and construction activities was fundamental to this complex project from the outset. Initially an outline overall framework was indicated in the contract specification by means of a schedule of key dates, together with the intended timing for release of drawings and information as the detail design proceeded.

37. This paralleled construction and was further complicated by the design input required from the nominated sub-contractors as they developed the detailed M&E systems and equipment within the requirements of the conceptual design and specifications. Subsequent analysis and equipment testing were then required to demonstrate seismic qualification, and contribute to production of the overall safety case.

38. In certain areas the contractor was required to develop the consultant designs to the level of detail required for construction and produce the requisite drawings for approval. Over 1,300 connection and fabrication drawings were required for the Shiplift structural steelwork alone.

39. Due account was taken of this in the programme, together with the contract requirement that method statements and QA documentation all had to be in place prior to commencement of any activity.

40. Co-ordination of these inter-related design processes and the various facets of multi-disciplined construction required sophisticated planning techniques. To manage and monitor this the contractor elected to use his main frame Artemis 9000 system operated from site based terminals.

This was adapted to operate multi-user multi-network format using hierarchal management codes to give four levels of output, controlled via a purpose written menu system. Progress measurement was by work achieved and against key dates on the critical path.

41. An early start to the marine piling operation was fundamental to the overall programme, and following contract award concentration centred on development and procurement of the extensive range of plant, temporary works and materials required. The programme for the NUB however dictated that this also started immediately. Prompt provision of the temporary site establishment to accommodate contractor's and consultants' staff and workforce presented a further priority.

42. The initial programme covered in detail only the principal structural elements, with the remainder being in outline, within the parameters of the key dates defined. Subsequent elements of the project were then incorporated into the programme as detail designs were finalised and released.

MARINE PILING

43. Early establishment of the piling activity was essential and presented one of the principal challenges of the project. In essence, within four months of contract award, and one month of access to site, it was necessary to design, procure, manufacture and establish an extensive range of equipment and temporary works.

Fig. 6 - Shiplift Piling During Installation

SHIPLIFT AND FINGER JETTY

44. This included jack-up barges and cranage; piling hammers and drills, including development to suit project specific requirements and noise restrictions; piling gates; pile bracing and temporary hard point; pile test rigs; the concrete batching plant; together with shipping arrangements for cement and aggregates; the pile fabrication facility and marine equipment for sea delivery to site.

45. The level of temporary works design required was extensive, particularly in respect of the pile bracing where computer modelling techniques were employed to analyse the developing 'structure' of piles, bracing and deck during the various phases of construction.

46. The choice of construction methods and equipment was based on review of the physical conditions, specification requirements and time restraints. Programme requirements were of the essence and initial considerations centred on the number of teams which would be required to execute the works within the required timescale.

47. The Finger Jetty piles could be accommodated by one piling crew, working sequentially out from the shore, utilising a crane running on a temporary crane deck, with deck and gates moving progressively forward on completed piles.

48. For the Shiplift however, time restraints did not permit a similar approach since the critical path ran through completion of both quays to meet the key date for Shiplift platform installation. An offshore start was necessary to achieve programme, together with a faster turn round time than achievable from a crane deck. This necessitated the use of either floating plant or jack-up barges, with the latter providing significant advantages.

49. To accommodate both quays simultaneously two jack-up barges were needed, together with a third jack-up barge to execute the remainder of the Shiplift piling. A worldwide procurement exercise confirmed a limited market for suitable equipment, but three appropriate jack-up barges were located. The Jay Robertson in New York; the Castoro, located off Ravenna in Italy, and the Buzzard from Amsterdam.

50. Following contract award the Jay Robertson and Castoro were brought to site on heavy lift ships, with the Buzzard being towed the shorter distance from Holland. Both the former were equipped with Manitowoc 4100 cranes on a ringer attachment and the Buzzard with a tracked Demag 2000 fitted to the barge in Amsterdam prior to delivery. The primary cranes were all fitted with a dual lift system to facilitate rotation of piles from the pile delivery barge from horizontal to vertical mode for installation in the piling gates. Secondary lifts were accommodated by means of a rough terrain mobile crane on each barge.

51. All of the 1,036 marine support piles were required to be driven to a prescribed set into the bedrock, and on the Shiplift all the 188 raking piles, plus a further 295 vertical piles were required to be installed in sockets in insitu concrete between 2m and 8.5m into the bedrock, with the piles driven through the fresh tremmied concrete to set in the bottom of the sockets.

52. Sockets were drilled using the pile as the drill casing, and were required to be under-reamed to one or two stages dependent on pile/socket type. It was not possible to under-ream right up to the toe of the pile and therefore a shoulder of rock remained to be broken off during the final drive into the socket, a process potentially aggravated by the presence of external shear rings.

53. To overcome the problem of pile instability in deep water with low overburden restraint, all piles were required to be supported by a system of temporary bracing, tied back to a firm support, until cast into the permanent deck. Working from land the bracing system was restrained by piles in substantial overburden depths plus the existing sheet piled seawall. The offshore start essential to the Shiplift programme necessitated installation of a temporary structure, or 'hard point', in 22m of water which could then provide an anchor for pile support. Ease and speed of construction was an essential design prerequisite for this and the chosen solution comprised eleven 1m diameter piles cast into drilled sockets in the bedrock with a rigid, two level, fabricated steel deck above low water level. Access was subsequently provided by mooring a 110m long pontoon between the shore and the hard point.

54. Numbers of hammers and drills were determined by the requirement to support four piling crews. This equipment was standarised to allow interchange between rigs and thus maintain flexibility. In determining the choice of piling hammer due account was taken of the sets to which these were required to be driven and the vulnerability of piles suddenly brought to a halt at rockhead level, particularly the longer thin walled raking piles. Diesel hammers were therefore discounted in favour of more controllable hydraulic hammers. Here there was a gap in the range on the market but the BSP HH7 provided an acceptable compromise, being sufficiently powerful to achieve the required set for all piles whilst limiting the risk of damage to the pile. At the Finger Jetty, where there was a significant depth of stiff boulder clay, some relief drilling was required to reach rockhead.

55. Analysis of sets was carried out using wave formula analysis initially based on estimated rock parameters and later confirmed by parameters derived from dynamic analysis.

56. Drill procurement was critical to the programme.

Four pile head rigs were required with under-reaming drill heads specifically developed and manufactured for the specified socket dimensions. The first complete rig was required on site in the first month of the contract to carry out 3 preliminary onshore tension pile tests.

57. It was elected to utilise Wirth drilling equipment, with PBA5 pile top assemblies. These relied on 'crowd' onto the pile tops to provide drilling pressure, derived from the weight of the assembly plus pile casing itself. To resist the torsion applied by the drill the pile was temporarily fixed to the piling frame, which also supported the weight of pile plus drill assembly up to 8m above the piling gate.

58. Evacuation of the drill cuttings, and the overlying overburden was by reverse circulation, fed from high pressure compressors. Due to the possibility of asbestos contamination of the overburden the air lifted arisings were discharged below water. Sampling of the cuttings was undertaken to ensure that sound rock had been encountered.

59. Concrete for the pile sockets was produced by the site batching plant and, where land access was not available, was loaded into skips for transport to the piling rigs by motorised barge. This was then fed to the pile sockets via a tremmie pipe before driving the pile to its final set into the base of the socket.

JETTY DECK CONSTRUCTION

60. Following piling the critical path passed through jetty deck construction. The jetties required a combination of formwork solutions to cater for their complex shape and variable soffit levels. Bottom slabs to the Finger Jetty and Berth 12 lent themselves to rolling forms, as did the duct sections to these areas where rolling tunnel forms were used to cast both duct walls and top slab integrally. Elsewhere a complex array of purpose built panels were used, which significantly extended the turnaround period per bay. The method of supporting the soffit forms was by utilising the pile bracing collars suspended by hanger bars from lugs welded to the pile caps, the collars being designed to fulfil both functions.

61. To limit shrinkage all concrete pours were restricted to a maximum dimension of 10m in any plane, adjacent bays had to be cast at least three days apart, and infill bays could not be concreted until adjacent bays were a minimum of 14 days old.

62. Efficient handling and storage of temporary and permanent materials presented a significant logistical challenge, particularly with regard to reinforcement handling. Due to the programme necessity for an offshore start, all materials had initially to be transported from shore on pontoons, with the decks serviced by pontoon mounted cranes, but later, as the piling and deck construction progressed, tower cranes were established

running on rails on the top slab. The Finger Jetty was simpler in this respect, with one tower crane being established on a temporary base on land initially, then moving onto rails on the deck. The Shiplift ultimately required deployment of three tower cranes, and at the peak of construction up to 11 cranes were fully engaged between piling and jetty deck construction within a relatively confined working space.

SUPERSTRUCTURE CONSTRUCTION

63. The structural steelwork frame to the Shiplift and the nine reinforced concrete towers represented the next critical path. Six of the reinforced concrete towers were slipformed with the other three being cast in sections in conventional manner.

64. Structural steelwork to the South Quay columns was erected from Berth 12 but the North Quay could only be sensibly accessed from the Shiplift platform once installed and suitably protected. This in turn led the critical path through to roof truss erection where, in order to reduce the overall duration of this activity, the individual members were bolted into sections outside the Shiplift, then transported onto the platform for assembly into complete roof truss sections. The whole roof was divided into four sections for assembly purposes with each quarter then being winched up into position and connected to the columns and neighbouring sections at high level.

65. With the superstructure steelwork frame in place construction works moved on to the relatively straightforward erection of cladding and reinforced concrete construction of the North Quay plant rooms and three storey South Quay accommodation block.

66. Finally M&E installation and building finishings concluded the facility.

SUMMARY

67. Overall the Shiplift package comprised the provision of a fully functional and essentially self-sufficient servicing facility for the Vanguard class submarine fleet.

68. Marine works constituted an essential element of this and a major challenge of civil engineering design and construction which were completed within the timescale dictated by the critical path. At the peak of the construction activity numbers of staff and workforce engaged on this element of the works alone exceeded 600 with a total of 1500 engaged in the Shiplift package as a whole.

69. The additional criteria of nuclear safety, seismic qualification and rigorous QA regime magnified both the extent of the challenge and the achievement of satisfactory conclusion of the end product.

The Faslane Syncrolift

G. A. STOKOE, President, and J. R. BERRY, Syncrolift Inc

SYNOPSIS. This paper, following-on from that given by VSEL, provides a second example of how investment in modern drydocking systems will provide immediate benefits to the Royal Navy, with the opportunity for low-cost flexibility in terms of decision-making on future shiprepair and maintenance considerations.

METHODS OF DRYDOCKING
1. If we consider drydocking techniques since 1805, then the SYNCROLIFT® (see Fig. 1) is a relative newcomer on the scene. Traditionally there were two ways to dock larger vessels, namely:-
    (a)   Graving Docks
    (b)   Floating Docks
2. However, both of these methods are by no means fully compatible with the requirements of modern production techniques, and are limited in terms of the number of vessels which can be handled at one time. They place the ship inside a space limited by two or more walls which severely restrict access.
3. Furthermore, expansion of capacity can only be achieved by adding more of the same; very rarely is it possible to 'grow' a Graving or Floating Dock in size as vessel requirements change. The only way to increase capacity is to build another Graving or Floating Dock, which will unavoidably have the same access restraints as its predecessor and preclude the utilization of modern techniques. The SYNCROLIFT® concept

Fig. 1 - A large SYNCROLIFT® (location San Pedro, California

Trident facilities. Thomas Telford, London, 1994

avoids these limitations and can readily be extended to provide increased lifting capacity.

4. Modern submarine building is based on modular construction techniques (see Fig. 2), whereby the outfitting of modules (in the form of hull sections) is continued for as long as possible before they (the hull sections) are joined together. This is very difficult to achieve efficiently in a large hole in the ground, or in a floating box. In days of yore, a submarine was built as a simple closed cylinder, which was then launched down a slipway into the water. Thereafter it took years for the shipbuilder to figure-out how to get the equipment inside, and labor had to work in extremely confined spaces! Once the vessel was in the water, it would require a drydocking should there be any work which necessitated work below the water-line. The SYNCROLIFT® concept permits modular construction techniques to be fully exploited in factory conditions and allows launch to be delayed until immediately prior to completion.

Fig. 2 - The SYNCROLIFT® concept permits modular construction techniques (location Adelaide, Australia)

5. Ship maintenance or shiprepair facilities are similarly limited in their flexibility of operation by the constraints of conventional drydocking methods.

6. Naval Architects design both surface ships and submarines to be supported by water - a totally compliant medium which provides buoyancy to vessels equivalent to the mass of water they displace. The buoyant forces vary along the length of a vessel, and are dependent on the area of displacement at each cross section. The inherent stiffness of the vessel's hull compensates for irregularities between these buoyant forces and the mass distribution of the vessel itself, and in so doing, the hull adopts a unique longitudinal profile.

7. Supporting a vessel out of water is an unnatural situation since the upward pressure of the water is replaced by point loads from the blocking.

8. The longitudinal profile of the blocking in traditional docking systems modifies the unique profile of the ship, when afloat, and introduces peaks to the longitudinal load distribution. The SYNCROLIFT® system with its articulated platform, on the other hand, more closely simulates the

"softer" support a vessel experiences while afloat and thereby reduces the undesirable loading peaks which are inherent in traditional docking systems.

THE SYNCROLIFT® CONCEPT

9. A SYNCROLIFT® is simply an elevator for lifting ships from the water to a level where work can be done in dry conditions on the platform itself or where they can be transferred to an on-shore work location. (Fig. 3)

10. An articulated, timber decked steel platform, equipped with a docking cradle, is lowered beneath the water to a depth at which a ship can be floated into position above the cradle (Fig. 4). The cradle is configured to support the ship's bottom profile. The platform is then elevated by a number of synchronized, electrically powered mechanical hoists which are positioned along both sides of the platform. The platform, carrying the cradle and the ship continues to elevate until it reaches transfer or working level, at which time a series of upper limit switches automatically stop further movement.

Fig. 3 - SYNCROLIFT® is an articulated timber-decked steel platform equipped with a docking cradle on rails (location Sabah, Malaysia)

Fig. 4 - The platform is lowered into the water...(location Oman)

11. At this point, the ship is ready to be worked on, or, in the case of SYNCROLIFTS® which are equipped with a transfer system, ready to be moved to an onshore dry berth for longer term work, leaving the elevating platform free to dock or undock other vessels (Fig. 5).

12. Platform lowering and raising movements are controlled by an operator at the control console located in an enclosed area alongside of the SYNCROLIFT®. This console includes the full complement of controls, indicator lights, alarms, load and platform position indicators necessary for the operator

to safely and efficiently control all aspects of drydocking and launching.

13. Located at each hoist are load cells which constantly monitor the loads at each hoist location and display their value on the control console. These readings are integrated into the safety system and provide assurance that overloads are not experienced.

BRIEF HISTORY

14. The original concept of the vertical shiplift goes back well over one hundred years and possibly further. The BRITISH JOURNAL OF ENGINEERING, in 1862, describes an 1800 ton hydraulic lift which was actually constructed in India. In the years between then and the mid-1950s, various primitive shiplifts were conceived and built, but all were either very small or proved to be very slow and unsafe. The shortcoming was always the means of coordination of the individual lifting machines. In small installations, limited success was achieved by using a common continuous shaft for all of the hoisting drums.

Fig. 5 - ...and then raised to transfer level for movement onshore (location Tandanor, Argentina)

15. The SYNCROLIFT® shiplift system was invented in 1954 by Raymond Pearlson, a Naval Architect, of Miami, Florida, U.S.A., the founder of the Pearlson Engineering Co. - the company which has pioneered shiplifting technology worldwide. Nowadays the company is known as Syncrolift, Inc. and is an important part of the Roll-Royce Industrial Power Group in the United States.

16. The development and application of the A.C. synchronous induction motor made the modern SYNCROLIFT® a viable concept. This motor can only operate at one speed regardless of its load. The motor speed is totally a function of the frequency of the A.C. power source. Thus the motors achieve the elusive goal of synchronization unsuccessfully sought by the early shiplift experimenters. Any number of identical motors operating from the same power source will behave as if they are mechanically coupled together. After years of research, SYNCROLIFT® has found no other system that more simply achieves the goal of mechanical synchronization.

17. The first SYNCROLIFT® was built in Miami, Florida in 1957 and is still operating today, using the same four motors, gears and controls installed over 37 years ago.

18. As the SYNCROLIFT® grew beyond simple four hoist lifts, it became evident that a continuous "rigid" platform

did not determinately distribute hull loads to each lift point. Adapting structural techniques which are commonly used in bridge construction and other structures, where emphasis on simplification of load distribution is desirable, SYNCROLIFT® developed the articulated platform concept, which it patented in 1963. As well as determinately distributing hull loads, this type of platform also reduces the undesirable loading peaks which occur with more traditional docking systems and rigid shiplift platforms. The importance of this feature to a successful and safe shiplift cannot be overstated.

19. During the 1960s, SYNCROLIFTS® for vessels of 100 tonnes to 10,000 deadweight tonnes were installed in countries from the Arctic to Equator. In most cases, on-shore transfer systems were installed in conjunction with the SYNCROLIFT® platform to permit the shipyard to accommodate several vessels at one time depending only upon the available land area.

20. During the 1970s, the requirement for even larger shiplifts became evident. In 1975, the SYNCROLIFT® at Las Palmas, Canary Islands, became the largest shiplift built to that time and the first to exceed 10,000 tonnes capacity. Its 172 metres long by 30 metres wide platform is operated by 64 SYNCROLIFT® hoists and the on-shore transfer system can handle up to 15 vessels at one time (Fig. 6).

Fig. 6 - This typical arrangement at Las Canarios shows 13 vessels docked by just one SYNCROLIFT® (location Las Palmas, Spain

21. The largest shiplift platform in the world was commissioned in 1984 at Todd Pacific Shipyards Corporation, Los Angeles Division. This SYNCROLIFT® has a platform 200 metres long by 32 metres wide, utilizes 100 SYNCROLIFT® hoists, and has a maximum lifting capacity of 22,000 tonnes.

22. The shiplift which now has the greatest lifting capacity is the subject of this paper. It was commissioned in 1993 at the Clyde Submarine Base, Faslane. The SYNCROLIFT® has a total lifting capacity of 25,000 tonnes, provided by 92 SYNCROLIFT® hoists, and a 165 metre long x 25 metre wide platform.

THE FASLANE DESIGN

23. This is the first shiplift to be specifically designed

to meet the safety standards established for handling nuclear submarines. The SYN-CROLIFT® concept has met the increased requirements of nuclear safety.

24. Initially the Design Specification defined fundamental requirements such as size parameters, applied loading, codes and standards, intended uses, interface requirements and constraints, speed of operation, standard of painting, environmental conditions and safety features.

Fig. 7 - HMS Vanguard rolling out for launch at VSEL Barrow (location Barrow, U.K.)

25. Subsequent revisions tended to increase the severity of these requirements and introduced further requirements, such as additional codes and standards, survival of failure scenarios, seismic criteria, blast resistance, new loading combinations, cradle requirements and design verification requirements, as the need for them was identified by the Safety Authorities.

26. A considerable number of finite element model analyses were necessary to justify features of the design which normally would be addressed by engineering judgement, or to justify aspects of the design which were not covered by the specified codes and standards.

27. The articulated platform consists of 46 transverse main beams which support the secondary steelwork, timber deck and rails. The main beams are assembled into 23 two-beam rigid modules weighing up to 200 tonnes each. Articulated sections of secondary steelwork span between the rigid modules upon which they are supported by rocker bearings. The platform articulation is not visibly evident because the synchronization of movement of the platform is so precise.

28. A series of 92 identical SYNCROLIFT® hoists (each rated at 387 tonnes lifting capacity, 400 tonnes static) are positioned along each side of the platform (i.e. 46 on each side), each supporting one end of a transverse main beam by wire rope reeved through sheaves mounted on the beam and at the hoist. The hoists are driven by an electric motor through a gearbox which drives a bull gear, which is integral to the wire rope drum. The drums are grooved to accept the full length of wire rope, so that only a single layer of rope is wrapped on the drum.

29. A further 4 hoists of 110 tonnes lifting capacity each, support two further main transverse beams at the offshore end of the platform. This additional platform module only

Fig. 8 - SYNCROLIFT® has many other applications like this Caisson lift for a large Breakwater project (location in Japan)

functions as a "Road-Bridge" section, for access between the piers in certain operational modes. It is not loaded by vessels during any drydocking condition.

30. The main platform is itself "split" longitudinally into two separate sections with 80 of the 92 hoists functioning to lift the vessel itself, whilst the outboard 12 hoist units support and actuate a separate 6-beam module which is called the "Rudder" Section. This section can be operated independently of the "Main" section to facilitate rudder removal; propeller maintenance or change-out; and/or shaft removal. This operation is conducted when the "Main" section of the platform is pinned at the upper level.

31. The SYNCROLIFT® platform therefore effectively comprises a composite of three sections which at certain times can be moved independently of each other. These are:-

    (a) Main Section   80 hoists
    (b) Rudder Section 12 hoists
    (c) Road Section   4 hoists
         TOTAL   96 hoists

32. It should be noted that for all docking and undocking operations, all sections operate synchronously as one unit. Conversely, at other times the system affords the Operator the flexibility to have all three sections at different levels simultaneously and to derive the benefits therefrom, which no other form of drydocking can achieve.

LOAD MONITORING SYSTEM

33. The dead end of each hoist's wire rope is terminated using a load cell. The load at each of these load cells is continuously displayed at the control console. This feature provides for safe lifting since it permits any unusual or unforeseen loading condition to be noted in time to avoid damage. Moreover, it is an important operating feature because it permits the Dockmaster to detect the initial point of ground and to have continuous docking data at his disposal during the entire docking cycle. The load cells and motor current sensing relays shut down all of the motors of the SYNCROLIFT® before motor slip can occur, or if the

platform is overloaded.

34. The control system (which was designed by NEI Control Systems of Gateshead, England) thus provides the Dockmaster with a whole range of information not only about the shiplift itself, but very importantly it also provides a profile of the load (i.e. the Submarine) which it is lifting, so that full attention can be paid to maximizing safety of the vessel and avoiding local concentrations of stress to the pressure hull. This is a unique facility afforded by the combination of the Load Monitoring system and the concept of platform articulation around which the platform is designed. No other form of shiplift design, or indeed of any other form of drydocking system, can provide the Operator with this facility.

35. The hoist motors are specially designed. As already mentioned, they are synchronous induction motors and their speed is determined by the cyclic input of the A.C. current. Thus, they will only operate at a single speed regardless of load. They achieve full speed within a few cycles when starting, so, when used in multiple installations such as the Faslane SYNCROLIFT®, they behave as if they were mechanically coupled together. It is the motor type and the limitation to a single layer of wire rope on the hoist drum which results in the synchronization of platform movement.

36. The electrical control system comprises the reversing motor contractors and control logic to operate the SYNCRO-LIFT®. Interlocks ensure all operations on all hoists are performed in the correct sequences. All safety related functions have at least one backup system available for redundancy. The control system monitors its own performance and constantly checks that all essential components are operating satisfactorily. It will stop the entire SYNCROLIFT® if an abnormality is detected. The control system also monitors the output from the two separate load monitoring systems and will similarly stop the SYNCROLIFT® if any abnormal loads are detected.

37. The SYNCROLIFT® cradle system supports all vessels while they are out of the water, whether on the SYNCROLIFT® platform or ashore (Fig. 9). The cradle wheels run on and are supported by a rail system which permits longitudinal movement relative to the platform. Individual cradles featuring

Fig. 9 - The SYNCROLIFT® cradle system supports vessels while they are out of the water (location Cartegena, Colombia)

either keel support blocks, bilge support blocks or both are assembled into a train of appropriate size and configuration for the vessel about to be docked. When the docking is complete the vessel can either remain on the platform or be moved ashore over the end of the platform.

38. A side transfer carriage, recessed in a pit such that the rails on its upper surface align with the longitudinal transfer rails at the shipyard level, is provided. The end transfer cradles carrying the ship are moved onto this carriage. The carriage can be moved laterally, carrying the ship and transfer cradle to align with the longitudinal rails of any of the shore berths. Once aligned, the ship can be moved onto the selected shore berth. Thus two or more vessels could in due course be "parked" adjacent to the one SYNCROLIFT® providing the ultimate opportunity for flexibility in the provision of future drydocking capacity at Faslane, and obviating the need for the significant additional Civil construction costs which would otherwise be incurred in the provision of new Graving or Floating docks.

INSTALLATION

39. The contract to procure, manufacture, install, test and commission the Faslane SYNCROLIFT® was signed in 1988 and awarded to Clarke Chapman Projects (formerly Clarke Chapman Syncrolift Division). The plan was to quickly procure and install the SYNCROLIFT® platform onto the jetties then being constructed so that its deck could be used as a work area by the main contractor to erect the enclosing building (Fig. 10). Once this building was weather tight the SYNCROLIFT® hoists, control and load monitoring systems would be installed, tested and commissioned while the main contractor was completing the outfitting of the building.

Fig. 10 - Platform installation underway at Faslane. Crane (in foreground) is placing a twin-beam section (location Faslane)

40. Installation, using the services of Clarke Chapman Engineering Services, generally went according to plan. The platform was fabricated and erection of it was completed in 1989. Manufacture and procurement of the hoists, control and load monitoring systems was then progressed while the building was being constructed. In 1991 the second phase of SYNCROLIFT® installation commenced and by 1993 the SYNCROLIFT® was ready for testing and commissioning. Full load testing

Fig. 11 - Transfer cradles with the SYNCROLIFT® Hall behind (location Faslane)

and operational trials were then completed. By the end of that year and following final commissioning, the SYNCROLIFT® was ready for use and handed over to the Royal Navy (Fig. 11).

NUCLEAR SAFETY JUSTIFICATION

41. All modern SYNCROLIFTS® contain a range of protective and safety "shut down" features which have been developed and refined over the years, as experience has increased, to protect the SYNCROLIFT® from overload, and to avoid local concentrations of load at any point around the pressure hull. For SYNCROLIFT® projects designed to dock conventionally-powered vessels, a Programmable Logic Controller is used. This constantly monitors its own performance as well as constantly checking that all operations are carried out in the correct sequences and that all essential components are operating satisfactorily. Hoist motor currents are monitored in addition to the thermal/magnetic motor protection also provided. The PLC also processes data from a Load Monitoring System and displays the load at each hoist on the Operator's Console. If the PLC detects any abnormalities it stops the entire SYNCROLIFT®.

42. However, the Faslane Safety Case advisors believed it would be impossible to justify the safety of Programmable Logic Controller software so the older electro-mechanical type control system was selected for the Faslane SYNCROLIFT®. This system was specially designed so that all operating circuits were monitored by additional circuits, such that should any component fail to function the system would fail safe and stop the entire SYNCROLIFT® by at least two separate means.

43. This results in about three times as many cable terminations, relays and other components as would be usual for this type of control system in a SYNCROLIFT® designed for conventionally-powered vessels. There is also about three times the length of electrical cable - about 30 kilometeres in total. While the control system meets reliability targets required for nuclear safety justification, there is an adverse effect on availability because of the increased possibility of malfunction due to the far greater number of components.

44. A number of further features to facilitate safety justification were added to the design at the request of the Safety Case advisors. A second brake of diverse design was added to each hoist. A second load monitoring system was added to provide additional means to stop the SYNCROLIFT® in the event of overload. The platform was designed so that it would continue to safely support the specified loading in the event that part of its structure had failed. It was subsequently strengthened to carry higher loads than it had been designed to accommodate when the possibility of such loading occurring was postulated by the Safety Case advisors.

45. The frequency and severity of materials testing was increased as a result of Safety Case defect assessment and crack propagation studies carried out on all components in the load paths. In many cases the additional non-destructive testing had be be performed at site since many components were already installed.

46. Finally normal Operating Procedures were restricted to simplify the safety justification of the control system and pre-operation checks taking about eight hours to execute were introduced to confirm the integrity of the controls each time the SYNCROLIFT® is used.

47. This SYNCROLIFT® is the only drydocking facility ever to be justified to the exacting nuclear safety standards which were applied to it. We believe no other form or type of drydocking has ever been able to meet such exacting requirements, nor been subject to scrutiny in the same depth or to the same level of detail (Fig. 12).

Fig. 12 - Inside the SYNCROLIFT® Hall. Part of the transfer system can be seen in the foreground (location Faslane)

INTERFACE WITH OTHER DESIGNERS

48. Liaison with the Civil engineering designers responsible for the shiplift jetties was essential throughout the design and construction phases, as the changing requirements for the SYNCROLIFT® modified the loading imposed on the jetties. Similar liaison with the designers of the electrical supply system and a whole host of other sub-designers and consultants, who contributed to the overall multi-

faceted CSB Faslane construction project, was necessary for similar reasons.

MAINTENANCE REQUIREMENTS
49. A SYNCROLIFT®, like any other form of drydocking device, requires a moderate level of preventative maintenance in order to provide a long, safe life and maximum service. Thirty-seven years of experience has proven that those shipyards which devote time and resources to such preventative maintenance minimize down time, and realize reliable and safe performance from the equipment.
50. The nuclear safety justification made for the Faslane SYNCROLIFT® by the safety advisors obviously required much more frequent, and more detailed in-service inspections than would normally be required for Commercial (i.e. non-Naval) shipyards. To our knowledge, no other form or type of drydocking has ever been able to meet such requirements, nor been subjected to examination of this level.
51. For the same reason some of the normal maintenance procedures are also carried out more frequently than normal and all maintenance tasks carried out must be precisely recorded to support the through-life safety justification. To achieve the necessary traceability, all components installed on site, as well as all spares provided, have unique part numbers assigned so their location and maintenance history can be tracked and verified. A computer data base with the appropriate software is used to manage all these records. Comprehensive Operation and Maintenance Manuals have been prepared to document all requirements.

THROUGH-LIFE SUPPORT
52. Syncrolift, Inc. has commenced providing the through-life design support which is needed to maintain the safety justification certification for ongoing nuclear related operations. The company has also been contracted by the Royal Navy to carry out all maintenance activities during the twelve month period following hand-over to them.
53. In addition to the provision of "hands-on" support in the form of Direct Maintenance, Syncrolift, Inc. has committed to provide Design Support throughout the life of the equipment. Such support incorporates the full gamut of through-life support from providing continuity of spares supply, to the provision of advice and services associated with any proposed changes by the Royal Navy in terms of new or existing vessel design, and or changes in operational procedures which may develop throughout the long life of this versatile equipment (Fig. 13).
54. In conclusion, it is appropriate to record a vote of thanks by the management and employees of Syncrolift, Inc. to the Ministry of Defense (Navy) and in particular, Commodore Treby RN, the officers and men of the Clyde Submarine Base in Scotland, and Mr. John Coles, Director and his staff at Defence Works (Strategic Systems), Bath,

FASLANE SYNCROLIFT

Fig. 13 - HMS Vanguard arrives at her new home.

England with whom we have been very proud to be associated. We would also like to acknowledge the role played by Mr. Partington and his team from Tarmac Black & Veatch (formerly PSA Projects) as Project Managers for both the Design Commission and the Supply of the SYN-CROLIFT® equipment on site.

©Copyright 1993 Syncrolift, Inc.
Two Datran Center
9130 South Dadeland Blvd.
Suite 102
Miami, Florida 33156-7848 U.S.A.
Telephone: 305-670-8800
Telefax: 305-670-9911

DATA SHEET	
CSB FASLANE SYNCROLIFT®	
Total Lifting Capacity	25,000 tonnes
Platform Size Overall:-	Length 176.8 metres Width 25.03 metres
Main Section:-	Length 143.4 metres Width 25.03 metres
Rudder Section:-	Length 21.5 metres Width 25.03 metres
Roadbridge Section:-	Length 11.9 metres Width 25.03 metres
Platform Weight	5,200 tonnes
Vertical Travel	23.0 metres
Lifting Speed (Main)	.022 metres/min
(Roadbridge)	1 metre/min
Hoists:- Main 92	387 tonnes lifting 400 tonnes static
Roadbridge 4	110 tonnes
Power Consumption	134 MVA

Seismically qualified jib cranes

N. P. HANCOCK, Chief Engineer, Stothert & Pitt, Clarke Chapman Ltd

SYNOPSIS

Stothert & Pitt have designed and manufactured three high integrity jib cranes now installed on Berths 10, 11 and 12 at the Clyde Submarine Base, Faslane, as part of the development of the Trident Submarine Support Facility.

The cranes, a 125 Tonnes lifting capacity Cantilever Crane and two 20 Tonnes lifting capacity level luffing Jib Cranes, have been designed to satisfy stringent safety requirements laid down by the Ministry of Defence.

Of critical importance was the requirement to seismically qualify both the crane structures and the crane electrical control equipment to ensure safe operation of the crane during the 1 in 100 year return seismic event and to ensure that the cranes remained safe, i.e. did not collapse or drop the supported hook load, during the 1 in 10,000 year return seismic event.

To achieve this Stothert & Pitt have conducted an exhaustive series of linear and non-linear computer Finite Element Analyses on both Crane Only and Combined Crane and Jetty models using 'state of the art' computer techniques.

The final crane designs have been successfully seismically qualified after a number of design and analysis iterations made necessary by the high seismic excitation levels applied to the cranes.

INTRODUCTION

1. Stothert & Pitt, a business unit of Clarke Chapman Ltd and part of the Rolls-Royce Industrial Power Group, have designed, manufactured and installed three high integrity jib cranes on Berths 10, 11 and 12 at the Clyde Submarine Base, Faslane, as part of the development of the Trident Submarine Support Facility.

2. The three cranes are all travelling electric cranes capable of 360° continuous slewing and comprise -

 (i) A 125 Tonnes lifting capacity Cantilever Crane installed on the Berth 10 and 11 Finger Jetty as shown in Figure 1.

(ii) A 20 Tonnes lifting capacity level luffing Jib Crane also installed on the Berth 10 and 11 Finger Jetty as shown in Figure 2.

(iii) Another 20 Tonnes lifting capacity level luffing Jib Crane installed on the Berth 12 Jetty adjacent to the Shiplift Building.

Figure 1- 125 Tonnes lifting capacity Cantilever Crane installed on the Berth 10 and 11 Finger Jetty

3. All three cranes have been designed to suit the handling of sensitive hazardous loads. The 125 Tonnes Cantilever Crane in particular has specified sensitive operations to perform including the handling of the Active Inert Missile System (AIMS) device and Submarine Ballast Cans.

4. A detailed Safety Justification Report for each crane has been prepared and submitted to the Safety Authorities in order to demonstrate compliance with the stringent safety requirements laid down by the Ministry of Defence Safety Consultants which include the following -

(i) Seismic qualification of the crane structures.
(ii) Seismic qualification of the crane safety related electrical equipment.
(iii) Structural and crane stability qualifications during specified exceptional loading conditions including extreme out-of-service wind and specified component failure conditions.

(iv) Materials of construction qualification to suit extreme temperature conditions.
(v) Reliability and Availability qualification.
(vi) Electrical equipment Electromagnetic Compatibility (E.M.C.) qualification.

5. This paper discusses the seismic qualification work performed on both the crane structures and also on the crane safety related electrical equipment. Particular emphasis will be given to the seismic qualification of the 125 Tonnes Cantilever Crane and the 20 Tonnes Jib Crane installed on the Berth 10 and 11 Finger Jetty.

Figure 2 - 20 Tonnes lifting capacity level luffing Jib Crane installed on the Berth 10 and 11 Finger Jetty

STOTHERT & PITT'S PREVIOUS SEISMIC EXPERIENCE

6. Stothert & Pitt's previous experience of the seismic qualification of Jib Cranes includes qualification work to varying specifications for cranes installed at the following locations -

(i) Royal Navy Dockyard, Devonport.
(ii) Royal Navy Dockyard, Rosyth.
(iii) Clyde Submarine Base, Faslane.
(iv) Royal Naval Armament Depot, Coulport.
(v) Vickers Shipbuilding and Engineering Ltd, Barrow.

7. The types of cranes supplied have included Cantilever Cranes of 80 to 125 Tonnes lifting capacity and level luffing Jib Cranes of 10 to 50 Tonnes lifting capacity.

FASLANE CRANES SEISMIC QUALIFICATION REQUIREMENTS

8. The cranes for Berths 10, 11 and 12 at CSB Faslane have been assessed against three specific levels of seismic excitation as described below. For ease of reference each seismic excitation level is described in terms of the zero period acceleration (ZPA) of the seismic horizontal response spectra at rockhead level.

   a. <u>0.05g ZPA reference seismic event</u>

      (i) The 0.05g reference seismic event corresponds to the 1 in 100 year return event level.

      (ii) The crane structures are required to withstand this level of seismic excitation as a Normal Working condition and require no remedial action.

      (iii) The crane safety related electrical equipment is required to maintain safe functionality during the 0.05g event. Sensitive equipment however is allowed to be reset in a 24 hour period.

   b. <u>0.2g ZPA reference seismic event</u>

      (i) The 0.2g reference seismic event corresponds to the 1 in 10,000 year return event level.

      (ii) The crane structures are to withstand this level of seismic excitation with specified minimum safety margins as an Ultimate Survival condition. The cranes are to remain structurally intact, must maintain the support of the hook load but need not necessarily be serviceable after the event.

      (iii) Any essential crane safety related electrical equipment is required to maintain safe functionality during the 0.2g event.

   c. <u>0.28g ZPA reference seismic event</u>

      (i) The 0.28g reference seismic event is analysed to prove that the 40% structural safety margin requirement for the 0.2g reference seismic event has been achieved.

      (ii) The cranes are to remain structurally intact at this level of seismic excitation and the support of the hook load must be maintained.

9. The seismic qualification of the crane structures in accordance with the above requirements was required to be achieved within stated constraints on the overall crane weight and on the maximum crane corner loads applied to the civil supporting structures as these structures were already designed and under construction.

THE DESIGN PROBLEM

10. Stothert and Pitt had therefore been set a difficult design problem -

(i) to design cranes to provide the required operating performances in terms of the loads to be lifted, hook radii, hoisting speeds, etc.

(ii) to qualify the cranes with respect to specified extreme conditions including seismicity.

(iii) to achieve the above without exceeding specified weight and corner load limitations.

11. Seismic qualification of cranes within such constraints is particularly difficult as the seismic excitation levels experienced at the base of the crane depend primarily on the dynamic characteristics of the structure supporting the crane, characteristics which are in general outwith the control of the crane designer. As a consequence, for cranes supported on civil structures, a unique seismic qualification is necessary for each individual installation as the seismic excitation experienced by the same crane design for the same rockhead seismic event will vary from jetty to jetty depending on the specific jetty dynamic characteristics.

SEISMIC QUALIFICATION OF THE FASLANE CRANE STRUCTURES

Qualification method

12. The structural seismic qualification of the cranes on Berths 10, 11 and 12 at CSB Faslane was achieved by conducting an exhaustive series of linear and non-linear computer Finite Element Analyses using the Nuclear Industry recognised analysis package 'ANSYS'.

13. Finite Element computer models of the cranes were developed and element plots of the models for the 125 Tonnes Cantilever Crane and 20 Tonnes Jib Crane on the Berth 10 and 11 Finger Jetty are presented in Figures 3 and 4 respectively.

14. Finite Element computer models of the combined cranes and jetty for the Berth 10 and 11 Finger Jetty were also developed and an element plot is presented in Figure 5. The combined models, which were based on a jetty model free-issued to Stothert and Pitt by the Project, can be seen to be very large and detailed. Computer run times for the analysis of these models was initially substantial and the use of powerful computer facilities was necessary in order to reduce the initial run times of in excess of 21 days down to more acceptable levels.

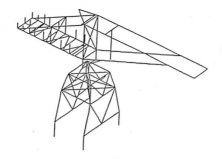

Figure 3 -   125 Tonnes Cantilever Crane Finite Element model

Figure 4 -   20 Tonnes Jib Crane Finite Element model

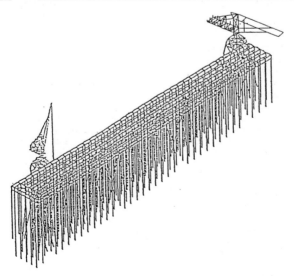

Figure 5 -   Combined Cranes and Jetty Finite Element model

15. The final seismic qualification method adopted for the Berth 10 and 11 Finger Jetty cranes involved the following types of seismic analyses -

(i) Linear response spectrum analyses.
(ii) Linear time history analysis.
(iii) Non-linear time history analysis.

16. Linear response spectrum analysis is an analysis in the frequency domain using, as input, dynamic acceleration response spectra at the base of the crane which define the levels of dynamic acceleration produced for different system response frequencies. In the case of the cranes, the spectra are secondary response spectra (SRS) as the cranes are 'secondary' systems mounted on the 'primary' system i.e. the jetty. The output produced from this analysis method is in the form of the maximum dynamic response of the system in terms of displacements, loads and stresses.

17. Linear time history analysis is an analysis in the time domain using, as input, acceleration or displacement time histories at the base of the crane. The output produced from this analysis method is in the form of time histories of the system displacements, loads and stresses throughout the duration of the analysed event. This analysis method is considered by Stothert and Pitt to produce more accurate results compared to response spectrum analysis but is considerably more onerous in terms of analysis time and cost.

18. Non-linear time history analysis is similar to linear time history analysis except that special computer software is used capable of analysing models with non-linear characteristics.

19. Structural analysis is normally conducted in the linear analysis domain where no discontinuities are considered and all components of the systems are considered to be able to react both compressive and tensile loads. However the structural seismic qualification of the cranes on Berths 10, 11 and 12 at CSB Faslane to the requirements and constraints of the Project Specification discussed earlier, proved to be very difficult due to the high levels of seismic excitation to which the cranes were subject. As a consequence, seismic qualification of the crane structures using linear analysis methods could not be achieved and the adoption of non-linear analysis techniques was necessary.

20. In the context of this particular seismic qualification, the main non-linearity considered was at the interface of the crane wheels and the rails on the jetty on which the crane runs. Linear analysis considers the crane wheels to be fixed to the rails and therefore able to react both tensile and compressive vertical loads. In reality this is not the case, the crane wheels can only react vertical compressive loads and uplift can occur at the wheels if no compressive load is present. This situation could occur if the seismic excitation to the crane were to be sufficiently high to cause momentary instability, resulting in a cyclic 'rocking' motion of the crane across the rails.

21. ANSYS 'Gap' elements were therefore incorporated into the crane models at the crane corners. These non-linear elements, if carefully introduced and assigned with appropriate parameters, enable the conditions which prevail at the crane wheels to be precisely modelled and allow uplift at the crane corners to be simulated. The non-linear crane models were therefore able to simulate possible 'rocking' motions of the crane whereas the linear crane models could not, as demonstrated in Figure 6.

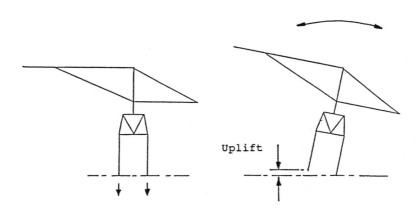

Linear models

o   Crane corners

Non-Linear models

o   Uplift at crane corners allowed

o   Crane able to 'rock' across the rails

Figure 6 - Differences between the linear and non-linear crane models

22. The adoption of this non-linear approach, whilst again being very onerous in terms of analysis time and cost, was very beneficial in substantially reducing the loadings and stress levels in the crane structures, due to the de-coupling effect created as the crane 'rocks'.

23. The concept of allowing the crane to 'rock' during the extreme seismic events might initially appear alarming. However, the magnitudes of the corner uplift predicted by the analyses were very small, amounting to a maximum of only 20mm during the 0.2g reference seismic event, which is insignificant when compared to the overall dimensions of the crane structures. This will be discussed in greater detail later.

24. The final seismic qualification methodology adopted and applied to each crane is reviewed in detail as follows –

Phase 1 – **Identification of critical crane configurations**

Initially 24 Crane Only models were developed covering the full crane operating envelope and including variants on hook load, hook radius and crane slew orientation.

These models were then subject to modal analysis to investigate natural frequencies and also to static analysis and seismic response spectrum analysis at the 0.2g reference seismic event to predict member stresses and crane reactions on the jetty.

The results from the analyses of the 24 Crane Only models were used to identify a reduced number of critical crane configurations for further detailed analysis.

The seismic input used for this stage of the analysis was the secondary response spectra at crane rail level that had been issued with the Project Specification.

Phase 2 – **Generation of seismic excitation data at the base of the cranes**

In order to generate seismic excitation data at the base of the cranes, linear time history analyses were performed on the Combined Crane and Jetty models at the 0.2g reference seismic event.

The seismic inputs to these analyses were free-issued acceleration time histories matched to the UK Hard Site Design Spectra and the outputs obtained were displacement and acceleration time histories at the base of the cranes.

Phase 3 – **Analyses of Crane Only models to produce the final qualification member loads and stresses**

Linear response spectrum analysis at the 0.05g reference seismic event and non-linear time history analysis at the 0.2g and 0.28g reference seismic events were performed on the selected critical Crane Only models to predict the final structural member loads and stresses and the crane reactions on the Jetty.

The seismic inputs to these analyses were secondary response spectra and displacement time histories at the base of the crane generated by the Phase 2 analysis of the Combined Crane and Jetty models.

Discussion of results

25. The structural seismic qualification of the cranes on Berths 10, 11 and 12 at CSB Faslane has been successfully achieved after a number of analysis and design iterations but has been difficult for the following reasons -

(i) The high levels of seismic excitation applied to the base of the cranes.

(ii) The Project Specification constraints on crane weight and maximum crane corner loads applied to the Jetty.

(iii) Unavoidable near resonant conditions existing between critical crane and jetty natural frequencies.

Modal analyses results

26. The results of the modal analyses of the 24 Crane Only models showed that, for the 125 Te Cantilever Crane, the most significant crane vibration was the transverse sway vibration, which is a horizontal translation of the crane mass above the truckframe portal level, as shown in Figure 7.

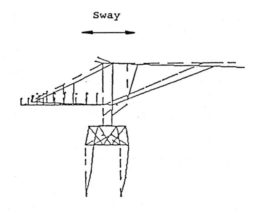

Figure 7 - 125Te Cantilever Crane transverse sway vibration

27. The frequency of this vibration for all 24 Crane Only models was found to fall on the peak of the jetty transverse secondary response spectra as shown in Figure 8. The crane transverse sway vibration was effectively in resonance with the jetty transverse sway vibration.

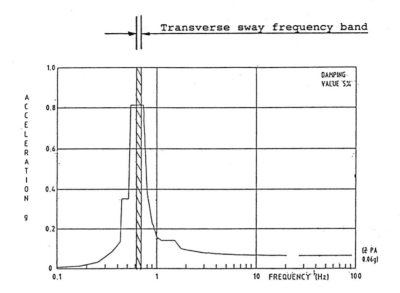

Figure 8 - 125Te Cantilever Crane transverse sway vibration frequency band

28. The critical parameter affecting the crane transverse sway vibration frequency was the height of the portal through the lower crane truckframe structure which, for the Finger Jetty cranes, was comparatively high. This parameter unfortunately could not be modified by Stothert and Pitt as it was a design constraint imposed by the Project Specification and was necessary to give the appropriate clearance over the buildings on the jetty over which the cranes travel.

29. Despite significant redesign and strengthening of the critical crane structures in an attempt to increase the crane transverse sway frequency it was not possible to totally de-tune the crane frequency from that of the jetty, and the crane structure therefore had to be designed for the high loads and stresses caused by this near resonance condition.

30. The modal analyses of the 24 Crane Only models of the 20Te Jib Crane on the Berths 10 and 11 Finger Jetty produced similar results to the 125Te Cantilever Crane. The crane transverse sway vibration frequency was again near resonance with that of the jetty and in addition, for the 20Te Jib Crane, near resonance conditions were also present for the main crane vertical vibrations. Again despite significant redesign and strengthening of the critical crane structures it was not possible to totally de-tune the crane frequencies from those of the jetty, and the 20Te crane structure also had to be designed for the high loads and stresses caused by the near resonance conditions.

Final crane linear and non-linear analysis results

31. The critical Crane Only models, selected from the original 24 models following the Phase 1 assessment, were subject to seismic analysis at each of the 0.05g, 0.2g and 0.28g reference seismic events, as described in Paragraph 24.

32. The analyses predicted that during the 0.2g and 0.28g events the cranes would experience 'rocking' motions across the rails producing uplift at the crane corners, as discussed in Paragraph 20. Figure 9 presents a typical time history plot of the predicted vertical displacement at the worst crane corner for the 125Te Cantilever Crane during the 0.2g reference seismic event.

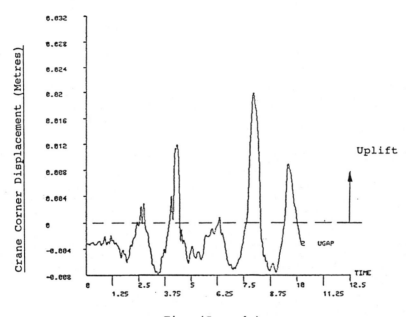

Figure 9 - Vertical displacement at crane corner for 125Te Cantilever Crane during 0.2g reference seismic event.

33. The crane can be seen to be subject to 5 rocking motions with a maximum predicted uplift at the crane corner of 20mm occurring approximately 7.5 seconds into the 10 second duration seismic event.

34. Table 1 presents a summary of the general crane motion predicted for each of the two Finger Jetty Cranes during the 0.2g and 0.28g reference seismic events in terms of the predicted maximum corner uplifts and the predicted number of crane rocking motions.

Table 1 - Summary of general crane motion during the 0.2g and 0.28g reference seismic events

Crane	0.2g Seismic Event		0.28g Seismic Event	
	Max Corner Uplift	No. Of crane rocking motions	Max Corner Uplift	No. of crane rocking motions
125Te Cantilever Crane	[mm] 20	5	[mm] 68	5
20Te Jib Crane (Berth 10 & 11)	7.5	2	31	4

35. It should be noted that the magnitudes of the corner uplifts predicted are very small when compared to the large overall dimensions of the crane structures. Nevertheless, the reductions in structural loads and stresses produced, when compared to the results for a crane modelled as fixed to the rails, are substantial. This is because, during uplift at the crane corners, the crane is effectively de-coupled from the seismic excitation being transmitted by the jetty.

36. The adoption of non-linear analysis techniques therefore to model the possible 'rocking' motion of the crane when subject to high seismic excitation proved to be very beneficial and was the critical factor in enabling the crane structures to be seismically qualified.

37. As previously mentioned, in order to achieve seismic qualification of the crane structures to the requirements and constraints imposed by the Project Specification, major redesign of large parts of the crane and a number of design and analysis iterations were necessary. As a general principle, in order for optimum cost-effective designs to be produced when seismic qualification is a requirement, it is essential to ensure the earliest possible liaison between the Crane Designer and the Supporting Structure Designer so that the two design expertise can be combined to remove any unnecessary constraints to the individual designs.

39. An indication of the degree of redesign necessary for the two Finger Jetty cranes can perhaps be given by a comparison between the overall weights of the final crane designs and the original crane designs being a 25% increase for the 125Te Cantilever Crane and a 40% increase for the 20Te Jib Crane.

SEISMIC QUALIFICATION OF THE FASLANE CRANES SAFETY RELATED ELECTRICAL EQUIPMENT

40. The seismic qualification method adopted for the safety related electrical equipment on the Faslane Cranes in accordance with the requirements of Paragraph 8 was primarily based on the generation of seismic secondary response spectra at the specific equipment locations for comparison with existing test spectra data for the equipment concerned.

41. The secondary response spectra were generated by analysis of the Crane Only models created for the structural seismic qualifications, modified to include greater detail in the areas of the electrical equipment houses.

42. Figure 10 presents a typical qualification showing the required response spectra (RRS) corresponding to the seismic excitation at the equipment location to be below the existing test response spectra (TRS) demonstrating successful qualification.

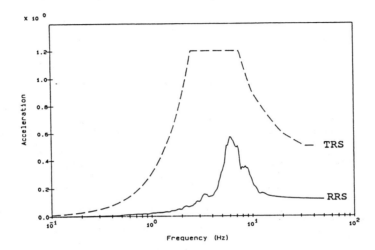

Figure 10 - Typical electrical equipment qualification spectra

43. The design of the electrical systems on the cranes is inherently failsafe by nature with all motion brakes being of the spring applied, powered off design such that loss of power results in all brakes being applied.

44. The cranes are fitted with seismic switches, set to operate at the 0.05g reference seismic event level, which cause the main power to the crane electrical systems to be removed, thus applying the motion brakes, by dropping out the main supply contactors.

45. The majority of electrical equipment was therefore only required to be seismically qualified for functionality during the 0.05g event with only the seismic switches, the main contactors and associated connecting equipment requiring qualification at the higher 0.2g event level.

46. This design approach enabled seismic qualification of the safety related electrical equipment to be achieved despite the high levels of seismic excitation at the equipment locations. The only components requiring specific seismic shaker table tests in order to achieve qualification were the main contactors. All other equipment was successfully qualified based on existing test data.

HV distribution and jetty support services

R. BALL, Principal Consultant, EurIng A. D. CAMERON, Yard Ltd, and H. A. CHERRY, Kennedy and Donkin Power, Consulting Engineers

SYNOPSIS This paper describes the support services which have been provided for TRIDENT submarines at the Jetties of the Clyde Submarine Base Faslane. The electrical supplies that are required by submarine nuclear reactors are outlined, particularly the means of ensuring that these supplies are of the integrity required by Safety Authorities to maintain supplies to what is, from time to time, the largest concentration of nuclear reactors in the UK. The paper also addresses the special mechanical services problems posed when a submarine is out of the water on the Shiplift. Detailed description of all the services is beyond the scope of the paper. The more interesting areas are covered in some depth, together with a summary of the reasoning behind the technical decisions taken.

INTRODUCTION

1. Following the decision that purpose built Shiplift and Finger Jetty Facilities were to be provided at the Clyde Submarine Base (CSB) Faslane to berth TRIDENT Submarines, studies were undertaken to establish the way in which the necessary jetty support services should be provided to these new facilities. It was concluded that over 20 year old services supplying the existing Southern Area Berths could not be extended and consequently that a new Northern Utilities Building (NUB) and associated Primary Intake Substation (PIS) were required. At the same time, it was decided that the NUB would replace the Southern Utilities Building (SUB) as the source of electrical supplies to the existing Southern Area Berths, which were the subject of a modernisation programme planned to run in parallel with the TRIDENT works programme. The Works, therefore, represented a significant change to the jetty support services at Faslane. Ensuring that the Works were carried out without prejudicing the day to day operational capability of the nuclear submarine base posed considerable design and installation challenges. In addition, the Shiplift posed particular mechanical service supply problems because of the need to maintain adequate support services to the submarines once lifted out of the water.

2. The Nuclear Safety Authorities demand that the supplies provided meet defined availability criteria against such hazards as seismic event, fire and

accident. Much of the detail design of the services has been driven by the requirement to demonstrate through the Nuclear Safety Case that these criteria are met. A very high level of design documentation, quality assurance and design verification has therefore been necessary throughout the design, installation, commissioning and testing phases of the work.

GENERAL JETTY SUPPORT SERVICE REQUIREMENTS

Electrical Requirements

3. When a nuclear submarine is berthed it normally shuts down its reactor and is therefore unable to generate power to circulate reactor coolant and maintain reactor instrumentation. The safety of the reactor is therefore dependent on a guaranteed supply of power from the shore.

4. It is thus a fundamental requirement of MOD's nuclear safety rules that any nuclear submarine must have high integrity electrical shore supplies available to it before it shuts down its reactor. Under most operational conditions, the jetty services have to provide a single PRIMARY electrical shore supply to the submarine, but be capable of providing, within a predefined time limit, a STANDBY supply, which is independent of the Primary supply with respect to its source. The requirement is illustrated in Figure 1(a). In other situations, when some of the submarine's own electrical plant is unavailable, the shore supply has also to provide an ALTERNATIVE supply. This has to be independent of the Primary supply with respect to its source and its distribution to the submarine. This requirement is illustrated in Figure 1(b). In particular, the nuclear safety rules specify that a 2 hour rated fire barrier should exist between Primary and Alternative supplies and associated control and protection cables.

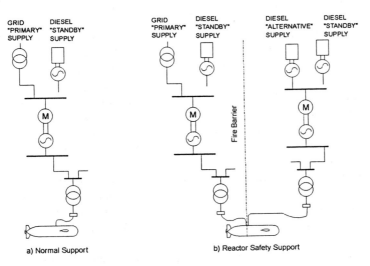

Fig 1 Basic Nuclear Submarine Electrical Supply Requirements

5. It should be noted that, although TRIDENT imposes a larger load on the electrical shore supply than previous nuclear submarines, its requirements are fundamentally no different to those of previous submarines. This has allowed well tried shore supply design solutions to be adopted.

6. TRIDENT, in common with all RN warships, has an electrical system that operates at 60Hz which is the NATO standard supply frequency. This means that frequency changers have to be used to derive 60Hz supplies from the 50Hz system. In addition, in common with previous classes of nuclear submarine, TRIDENT requires a variable, low voltage dc shore supply to support the reactor under certain operational conditions.

Mechanical Requirements

7. When alongside at the Jetties, TRIDENT is connected to mechanical services, notably potable water, diesel fuel, compressed air, nitrogen systems, domestic waste and industrial effluent. These services do not require to be continuously connected for the safety of the nuclear reactors. There are, however, two water supply services that are essential to the safety and support of the reactors especially when the submarine is raised on the Shiplift. The innovative design solutions adopted to overcome the cooling water problems imposed by the Shiplift are described later in this paper.

HIGH VOLTAGE SYSTEM ARRANGEMENT

8. The arrangement of the electrical distribution system is as shown in Figure 2. The distribution of 60Hz power required by the submarine is carried out at 6.6kV. This voltage was chosen to suit existing 60Hz distribution rings at CSB Faslane. Distribution of 50Hz power for all other requirements is undertaken at 11kV. The elements of the system are described in the following sections of this paper.

PRIMARY INTAKE SUBSTATION

9. The Primary Intake Substation (PIS) is the point at which Grid derived supplies are transformed to 11kV. The two PIS transformers are each rated at 15/30 MVA ONAN/OFAF, and are fitted with an on-load high speed, resistance tap-changer. The transformer output feeders are connected by cables to a 5 panel Vacuum Circuit Breaker (VCB) 11kV switchboard which in turn supplies the NUB via two cable feeders. In nuclear safety terms, the two PIS transformers are derived from a single source of supply, and so cannot be used to provide Primary and Alternative Supplies.

NORTHERN UTILITIES BUILDING

10. The NUB houses the high voltage distribution switchgear, diesel alternators, frequency changers, and the main and emergency control rooms. The building is arranged in four quadrants separated by fire barriers (see Figure 3) to meet with the nuclear safety segregation requirements.

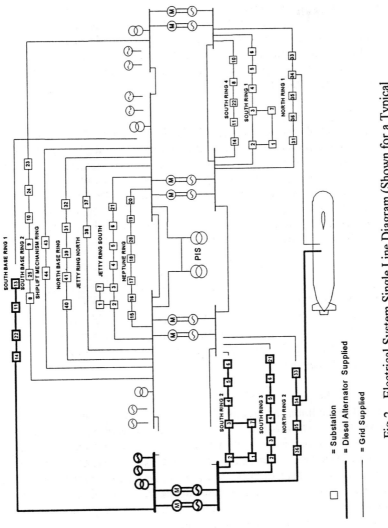

Fig.2 Electrical System Single Line Diagram (Shown for a Typical Reactor Safety Mode Configuration)

Diesel Alternators

11. The NUB contains eight 11kV, 50Hz, 3.1MW diesel alternators, arranged in pairs, with each pair housed in one of the segregated machine halls and connected to its own 11kV switchboard quadrant. The rating of the sets was determined such that six out the eight sets would meet a defined maximum loading condition (assuming that one set had failed and one was under maintenance), and that operation of any two sets would meet the minimum nuclear safety load. The alternators are air cooled, single bearing, brushless machines with a permanent magnet pilot exciter and salient poles. The diesel engines are water cooled, with the cooling effected via seawater heat exchangers.

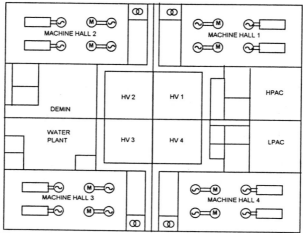

Fig.3 Arrangement of NUB Into Quadrants

Frequency Changers

12. The NUB contains eight 2.2MW 50/60Hz rotating frequency changers arranged in pairs, with each pair housed in one of the segregated machine halls, and connected to its own 11kV 50Hz and 6.6kV 60Hz switchboard sections. The rating of the sets was determined such that six out of the eight sets would meet the anticipated maximum 60Hz load (assuming one under maintenance and one failed) and that operation of any two would meet the minimum nuclear safety load.

13. The frequency changers are of the self ventilated, revolving field, salient pole, brushless type, provided with full interconnected damper windings in the rotor pole faces. The synchronous motors are suitable for operation at both leading and lagging power factors and can act as synchronous compensators to improve the overall power factor of the CSB load. A frequency changer is started as an induction motor utilising the rotor damper windings and an 3.5 MVA autotransformer starter.

14. The 6.6kV alternators are fitted with racking gear to allow the phase angle of the output voltage to be adjusted to permit incoming machines to be synchronised with already running machines. The racking gear is normally motor driven under the control of automatic synchronising equipment.

High Voltage Switchgear

15. The same type of vacuum switchgear is used in all NUB 11kV and 6.6kV switchboards and also for the frequency changer starters, in the PIS, and in the ring substations described later in this paper. 1600A, 1250A and 630A units are used, depending upon the duty concerned. The VCBs are rated to interrupt 25kA at 50Hz and 60Hz. Surge suppressors are fitted at the 11kV switchboards to suppress any overvoltages caused by VCB switching operations.

Control

16. Normal control of the NUB and associated distribution systems is by use of a computer screen based Supervisory, Control and Data Acquisition (SCADA) system from the Central Control Room (CCR). Redundancy is built into the SCADA system by using a Primary and Alternative philosophy. In the event of total SCADA failure, a separate Emergency Control Room (ECR) is provided and this is based on hardwired technology. The ECR is separated from the CCR by a 2 hour rated fire barrier.

17. The system is under operator control at all times but to limit operator workload, functions such as starting machines, synchronising and load sharing are all carried out automatically once initiated by the operator.

High Voltage Distribution

18. The HV Switchgear in the NUB is arranged as 11kV and 6.6kV single busbars split into four physically separate, but interconnected sections as shown in Figure 2. The required isolation between quadrants is achieved by the use of bus section VCBs at both ends of the cable that connects the busbars in different NUB quadrants. Supplies to local NUB auxiliary supply transformers are taken from the 11kV system, with one such supply from each quadrant.

19. Distribution of 11kV 50Hz electrical power to outlying substations is via seven ring main cable circuits. Generally, these supplies are transformed at the various ring substations to low voltage ac supplies for local use. However, the 11kV Jetty Rings North and South also supply the transformer rectifiers used to provide the dc jetty supplies required by the submarines. The connections of the two South Base rings are arranged to allow them to be used for load balancing purposes when the diesel alternators are in use.

20. Distribution of 6.6kV 60Hz electrical power to outlying substations is via six ring main cable circuits. These supplies are transformed at the ring substations to provide 450V 60Hz supplies to submarines at the jetties, to nuclear submarine related services and to other key 60Hz loads. Each berth is

supplied from at least two independent rings supplied from different pairs of 6.6kV, 60Hz busbar quadrants. This arrangement allows the required segregation between Primary and Alternative supplies to be achieved.

System Operating Modes

21. Under normal operating conditions i.e. when no berthed submarine has a requirement for "Reactor Safety Supplies", all 11kV and 6.6kV quadrant interconnectors are closed and all rings are operated as closed rings. The Grid provides the normal "PRIMARY" source of supply, but if the grid supply is unavailable, the diesel alternators provide the backup "STANDBY" supply.

22. The provision of Reactor Safety Supplies is achieved by splitting the electrical system into two, with one part, as normal supplied by the Grid, with the other part supplied by diesel alternators. Figure 2 shows a typical system split. The quadrant design of the NUB described above, and the design of the distribution systems detailed later in this paper ensure that these two supply systems are totally independent and are totally separated by, at least, a 2 hour rated fire barrier.

HIGH VOLTAGE DISTRIBUTION

Cables

23. All new 6.6kV and 11kV cables are 3c x 185mm^2 XLPE/CTS/LSF/SWA/LSF. This cable is also used in the jetty substations to supply the transformers and transformer rectifier units that supply the jetty ac and dc services.

Cable Ducts, Drawpits and Distribution Chambers

24. The high voltage distribution cables (and associated pilot protection cables and SCADA cables) are routed from the NUB to the Shiplift and Finger Jetty Facilities in underground fire clay cable ducts. The fire barriers provided by these ducts are maintained where the cables pass through cable drawpits.

25. The cable duct runs are seismically qualified to ensure that the electrical supplies are not disrupted by the defined seismic event. This aspect of the design required special attention at the Distribution Chambers which form the interface between the Shiplift and Finger Jetty Facilities and the Landward Infrastructure (see Figure 4). On the landward side, the cable ducts are routed in the loose material used to infill the space behind the new sea wall and original firm ground. In this region, the ducts are located in a concrete cable trough which, like the distribution chamber itself, is piled onto bedrock. Flexibility in the duct run between these two structures following a seismic event is provided by the relative angular movement between the individual lengths of fire clay duct. On the seaward side, a "seismic gap" is provided between the jetty structure and the distribution chamber structure. This gap allows the two structures to move relative to each other. All cables that cross

this gap are provided with sufficient extra free length at the gap to ensure the structures can move without damaging the cables.

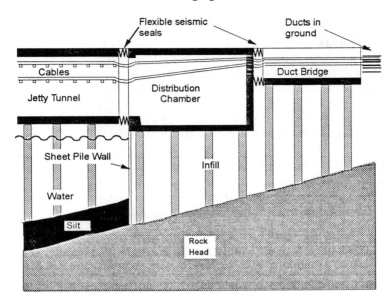

Fig.4 Seismic Interfaces at Distribution Chambers

26. Within the Distribution Chambers, the necessary fire barrier segregation between the Primary and Alternative cables is provided by encasing the cables in a set of ducts built out of proprietary fire barrier material. This also provides the necessary separation from the mechanical services which also pass through the Distribution Chambers.

Shiplift and Finger Jetty Service Tunnels

27. The Shiplift and Finger Jetty Facilities are provided with tunnels through which the mechanical and electrical services are routed. The necessary fire barrier between the high voltage supply cables is achieved by routing cables down different tunnels. In this way, the Primary and Alternative supply rings are separated from each other, as are the go and return legs of individual rings, and the mechanical services from all electrical services. Within the tunnels, the cables are fixed to seismically qualified cable racking. Fire barriers are installed at intervals along the lengths of the tunnels to reduce the spread of fire, smoke and toxic gases, and all cable penetrations are sealed for the same reasons. An automatic, zoned fire detection system is installed to provide immediate warning of any problems.

SUBMARINE JETTY SUPPLIES

Berth Substations

28. The arrangements within the jetties used to provide the submarine with electrical supplies are shown in Figure 5. The low voltage ac and dc supplies are provided by regulating transformers and transformer rectifier units respectively. These units are located in separate plant rooms in the Facility buildings above the tunnels and are supplied from separate high voltage switchrooms also in the Facility buildings. The plant and switchrooms are constructed with walls having at least a 2 hour fire barrier rating, thus ensuring that the nuclear safety segregation requirements are met. The high voltage supply rings are routed from the service tunnels directly into the switchrooms above the tunnels.

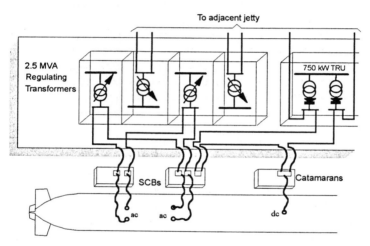

Fig.5 Diagrammatic Arrangement of Electrical Shore Supplies

29. The low voltage ac supplies are provided by 2.5MVA 6.6kV/440V self regulating transformers The voltage at the shore connection boxes is sensed and is automatically adjusted by the transformer to ensure that the voltage on the submarine busbars remains within its required limits. The two 2000A air circuit breakers are enclosed within a panel attached to the side of the transformer.

30. The low voltage dc supplies are provided by 750kW 11kV/185-750V dc transformer rectifiers. A control cable is connected to the submarine to allow the submarine's personnel to adjust the dc voltage as required to suit operational conditions. As with the ac transformers, the low voltage 1200A air circuit breakers are contained within a panel attached to the side of the transformer rectifier unit.

Submarine Overside Supplies

31. The low voltage ac and dc cables are routed in the electrical service tunnels from the transformer or transformer rectifiers to the submarine. Cables associated with one nuclear supply route are at all times separated from those of other possible supply routes. In most cases, this separation is achieved by running the cables along separate tunnels; in other instances, where cables from one supply have to cross the tunnel associated with the other supply, the crossovers are totally enclosed in a barrier constructed out of proprietary fire barrier material. The low voltage cables from the substations are of XLPE/LSF/SWA/LSF construction. At the jetty edge, the cables are taken to junction boxes in which the cable type changes to EPR/CSP in order to obtain the necessary flexibility to make the final connection to the shore connection box (SCB) which is mounted on the jetty catamaran.

32. Three catamarans are attached to each jetty as shown on Figure 5. Each catamaran is arranged to rise and fall freely with the tide, but is restricted in horizontal movement along and away from the jetty. The flexible cables are routed on flexible cable handling chains onto the catamaran and then enter the underside of the respective SCB. The mechanical design of the support chains ensures that the cables for each feeder circuit and their associated control and protection cables fixed to the support chains are separated from those of other circuits in accordance with the segregation principles. Electrical service booms positioned near the edge of the catamaran provide support for the flexible cables which are routed from the SCB to the submarine.

SYNCROLIFT SUPPLIES

33. The electrical power required by the Syncrolift Shiplift Mechanism is supplied by the 11kV 50Hz Shiplift Mechanism Ring which feeds two substations in the Shiplift Building (see Figure 6). One switchboard is arranged to supply the power requirements of the three Motor Control Centres (MCC) associated with the North Quay motors and the other similarly supplies the South Quay motors. Each two section switchboard is supplied by two 1.6MVA cast resin transformers; interlocks ensure that the transformers cannot be paralleled. Each MCC is connected to one section of the distribution switchboard with an alternate supply derived from the other section of the board. Although the supplies to the Syncrolift are not, in themselves, nuclear safety related (because loss of supplies presents no hazards to the submarine). They have been designed using the same segregation principles to ensure that the required availability targets are met.

34. The Shiplift Mechanism Ring also provides a supply to the sea water cooling pumps at the Shiplift. This supply is described in the next section of this paper.

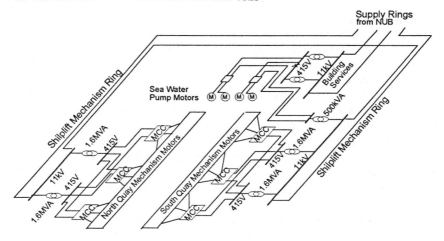

Fig.6 Shiplift Mechanism and Seawater Pump Supplies

REACTOR COOLING ON SHIPLIFT

High Pressure Decay Heat Removal

35. The preferred method of maintaining reactor cool-down whilst the submarines are in port is a system called High Pressure Decay Heat Removal (HPDHR). The reactor primary cooling water is passed directly from the reactor core through a cooling coil housed in an enclosed tank within the submarine. This is normally cooled by fresh water with its flow forced by the pressure in the fresh water system.

36. Whilst on the Shiplift platform, the necessary fresh water supply is transferred to the submarine by means of an articulated marine loading arm. This arm is similar to those used to carry electrical supply cables to the submarine when it is on the Shiplift. The arm, based on similar types used throughout the world for tanker loading berths, was rigorously analysed to ensure that the design would meet the seismic requirements for ruggedness of construction and reliability and survivability of the hydraulic operating and control components. The arm can be moved under hydraulic control to the submarine floating at the lowest tide condition in the Facility and after connection to the casing the hydraulic power pack and controls are switched off. Accurate tracking of the submarine as it rises on the platform is achieved by a slightly positive counterbalance which tends to raise the arm.

37. Although all elements of this cooling within the Shiplift Facility are qualified for hazards including the seismic event, the fresh water supplies outside of the base area were unqualifiable and, in the event of loss of the fresh water supply due to a seismic or other event, provision is made to allow change over to a qualified sea water supply.

Sea Water Cooling

38. Sea water is the normal cooling medium available to the submarine when afloat and is taken in via a number of inlet hull valves on the bottom or sides of the vessel. Continued availability of the seawater supply to the submarine when on the Shiplift is thus vital.

39. The Shiplift sea water supplies are obtained from 4 in number submerged bore hole pumps suspended below the Shiplift deck from the two independent sea water pump rooms. The pumps are electrically driven and have high integrity duplicate supplies as described below. Each pump can deliver $400m^3$/hr into a double ring main system serving both sides of the Shiplift. One pump could meet the normal requirement for cooling; however, it is intended that the supply should be operated with two pumps running, in a split configuration, guaranteeing an uninterruptible supply of cooling water.

40. The novel feature of the design solution is again the method of connection to the vessel which must cater for movement as the submarine is raised or lowered on the platform. After extensive evaluation of many alternatives, the Facility was fitted with a number of 3m diameter hose reels carrying 75mm diameter reinforced rubber hoses. These are reeled out to be connected to the submarine whilst in position prior to being raised on the platform and the hoses are then wound in as the vessel rises.

41. The final connection of the hoses to the submarine is made by means of instantaneous couplings which plug into flooding bonnets that are fitted by divers before the submarine is moved into the dock. The resulting system is relatively simple using robust, easily maintained or replaceable components that are flexible in use.

Sea Water Pump Electrical Supplies

42. The electrical supplies to the sea water cooling pumps referred to above are taken from two 2 section switchboards located in separated pumphouse switchrooms as shown in Figure 6. Each switchboard is provided with two supply and one bus-section switches which are mechanically key interlocked on a two out of three basis. One section of each switchboard is supplied from the 415V Building Services switchboard, which is supplied by North Base Ring, and the other switchboard section gets its supply from the Shiplift Mechanism Ring. These supplies are segregated throughout their length from the NUB, in a similar manner to the Submarine's overside supplies, to ensure that there is no risk of common mode failure leading to an inability to supply power to the pump motors. The Shiplift Mechanism Ring supply route is seismically qualified thus ensuring that supplies can be maintained following a seismic event.

ELECTRICAL PROTECTION

43. Protection is fitted to ensure that faults are removed as quickly as possible in order to maintain supplies to the remainder of the system. The electrical fault protection scheme for the primary electrical plant comprises main unit protection schemes (Translay Pilot wire, Busbar and machine differential schemes, Restricted earth fault schemes) with back-up provided by overcurrent and earth fault schemes. By their nature, these primary schemes provide good discrimination with fast clearance times by isolating the zone in question. No unit protection is fitted to the ring substation busbars; acceptable protection against the very unlikely faults on these busbars is provided by the ring end feeders.

44. The discrimination of the back-up system is not, and cannot be, as good as that for the primary schemes because clearance times for the back-up protection are inherently longer than those for the primary protection. However, the clearance times have been constrained to be well within the damage curves for equipment and plant. The schemes have been designed to provide discrimination when operating under either Grid or diesel alternator supply sources.

EARTHING SYSTEMS

45. The 11kV neutral of each PIS transformer is grounded via a 25 ohm metal grid type resistor. The 6.6kV 60Hz power system is earthed via 10 ohm resistors in the neutral-to-earth path at the alternator end of each frequency changer.

46. The secondary windings of the regulating transformers and the transformer rectifier units that provide jetty supplies to the submarines are not earthed. This philosophy follows that adopted for the submarines' electrical systems and allows for a single earth fault to be tolerated without causing loss of supplies to essential services.

47. The earthing network in the TRIDENT Facilities comprises a ring main of 240mm^2 copper conductor which is connected to the tubular steel piling of the jetty structures in two locations. All metalwork in the substations is connected by suitably sized copper conductors to solid copper earth bars which are, in turn, connected to the earth ring conductor. The armour and screens of the 11kV and 6.6kV ring main cables are solidly connected to the HV equipment earth bars in the substations and in the NUB, and so act as additional main earthing conductors between the substations and the NUB. The design adopted ensures that the return path for any earth fault does not rely on true earth but comprises a metallic conductor. All parts of the earthing network which may be required to carry power system fault current are capable of passing the prospective fault current in the respective part of the system for three seconds, by which time the fault will be removed from the system by the operation of the protective gear.

SYSTEM ANALYSES

Fault Level Studies
48. A comprehensive fault level analysis of the CSB Faslane power system was carried out using a detailed system computer model. The Interactive Power System Analysis (IPSA) package developed at UMIST was used for the studies. A number of configurations of both 50Hz and 60Hz were analysed. The maximum realistic fault levels at the NUB 11kV busbars, when Grid supplies and diesel alternators are in parallel when restoring suppliers after a Grid failure, were found to be 16kA. The absolute maximum fault levels, with all possible machines connected to the 11kV busbars was found to be 22kA. Estimation of fault levels on the low voltage jetty supplies to the submarine required careful assessment of contributions from power sources on the submarine, including its battery since, in some instances, these sources provided greater fault currents than the shore supply system.

Transient Studies
49. The IPSA power systems analysis package was used to study the dynamic performance of the system when subjected to the following sources of system transients:
- Starting of frequency changers
- Operation of the Shiplift Platform
- Major Short Circuits
- Splitting of the NUB 11kV busbars following a fault
- Loss of diesel alternators
- Loss of frequency changers
- Transients on the Grid supply system
- Starting of large loads on the submarine

50. The results of the studies were used to assess the performance of the protection systems and to confirm that voltage and frequency transients at the submarine supply busbars remained within their specified limits. The studies showed that pole slip protection was required on the frequency changer motors to protect them from the effects of large system instabilities that could occur in some circumstances. Trials were undertaken on a similar facility at Rosyth Dockyard to demonstrate that the Shiplift would, as predicted, regenerate power when lowering, and IPSA studies were used to determine the effect of Shiplift starting transients. Operating configuration modes have been developed to ensure that the diesels are not subjected to reverse power during Shiplift operation.

Harmonic Studies
51. The main sources of harmonic distortion on the CSB system are the transformer rectifier units supplying dc power to: vessels alongside, cathodic protection schemes, static inverters and crane drive systems. To ensure that the quality of supply to the submarines is in accordance with the performance

requirements, a policy decision was made by MOD not to feed the dc shore supply rectifiers from the 6.6kV 60Hz system.

52. Initial work was carried out to measure and analyse the harmonics on the then existing 11kV 50Hz system at CSB. This gave information to help produce and verify a computer model used to predict the harmonic performance of the new system. The results showed that harmonic levels would be within the required limits for both Grid supplied and diesel alternator supplied configurations.

CONSTRUCTION

53. The installation of the mechanical and electrical Works was carried out by nominated subcontractors under the main contractor for the Shiplift Package, Trafalgar House (Major Projects) Limited. The Works in the NUB and on the SCADA system were supervised by Kennedy & Donkin Limited, and the remainder including the distribution systems and the works in the Finger Jetty and Shiplift by YARD. Both consultants were delegated substantial powers by the Designated Lead Consultant, Babtie, Shaw & Morton. The M&E Works commenced in late 1987 when the civil works were well advanced and were substantially completed in sections between September 1992 and November 1993. A significant feature of the installation was the co-ordination of the mechanical and electrical building and civil works, and the planning of the installation of larger items of plant.

54. The services and their appropriate support systems were the subject of a rigorous QA and testing regime. In addition, those services which had nuclear safety requirements were subjected to test procedures drafted by a specialist testing organisation. On completion of the installation, testing and commissioning of the various items of plant and services, a system testing programme was undertaken to prove the adequacy of the design of the complete systems.

CONCLUSION

55. The HV Distribution system and Electrical and Mechanical Jetty Support Services described in this paper have contributed towards one of the largest infrastructure developments in Europe in recent years. The need to support the TRIDENT submarine has required these services to be designed, installed, tested and commissioned to exacting nuclear standards. The Works have been carried out without interfering with the day to day operational capability of the Clyde Submarine Base.

The floating explosives handling jetty

J. R. WARMINGTON, Assistant Director, J. P. ABELL, Senior Engineer, and M. W. PINKNEY, Assistant Director, Rendel Palmer & Tritton

SYNOPSIS The Explosives Handling Jetty, (EHJ), at RNAD Coulport provides Trident submarine weapons handling facilities within a covered berth. The "U" shaped post-tensioned concrete hull supports a large enclosure building with two high integrity electric overhead travelling cranes and contains mechanical and electrical services. The berth is closed by a vertical left vessel door and caisson gate. A flexible mooring holds the EHJ on station with access via two bridges. The EHJ was constructed in a dry dock and, when substantially complete, towed to Coulport, moored, and commissioned. At some 85,000 tonnes it is one of the largest floating concrete structures in the world.

GENERAL INTRODUCTION

1. As a major component of the facilities at Faslane and Coulport to support the new Vanguard class submarines, currently under construction at Barrow, the Ministry of Defence, (MoD), identified the need for a new covered berth at the Royal Naval Armaments Depot, (RNAD), Coulport on Loch Long. Consequently a Brief of requirements was produced by the MoD for action by their procurement agency, the Property Services Agency, (PSA).

2. A number of development studies were initiated by the PSA including environmental issues. Options considered were a variety of fixed berths, a floating berth and an underground berth cut into the foreshore rock. The location of the new jetty was to be north of the existing Polaris jetty, at a radius from Ardentinny governed by non-nuclear blast regulations. At this location the lochside was in a virginal state, requiring the construction of access roads and a support area; see Paper 13.

3. As a result of these studies a decision was taken, based on cost projections, to proceed with the design of a floating jetty. At the chosen location, a fixed berth posed great difficulties in construction terms. The depth of water at the outer berth is about 80 metres and the slope of the bed, which is rock with negligible overburden, is of the order of 40°, rendering a piled solution extremely expensive. As there were strong opposing political lobbies for concrete and steel as the hull construction medium, a further decision was therefore taken to produce alternative tenders. In the end, concrete was the chosen construction medium.

4. At this stage, (mid 1984), the PSA awarded a commission to BAeSEMA, (YARD), for the design and supervision of the Mechanical and

Electrical, (M&E), works, including cranes. A commission to Rendel Palmer & Tritton, (RPT), followed in November of that year for the design and supervision of the Building and Civil Engineering, (B&CE), elements of the EHJ. plus the role of Lead Consultant; to coordinate the work of the M&E Consultants, the Architectural Consultants, (Building Design Partnership), and the Project Quantity Surveyors, (George Corderoy & Co.).

5. The foregoing sets the scene for the design activities of RPT and YARD in particular. Following the MoD Brief and the study work already carried out by the PSA, the Design Team produced a Preliminary Sketch Plan in 1985, followed by a Final Sketch Plan in 1986. These plans were essentially detailed feasibility studies outlining the proposed configuration, including, inter alia, proposals covering construction methods, contractual strategy, programme and cost estimates.

6. The general arrangement of the EHJ as designed is shown in Figs 1 and 2 and has been briefly described in the synopsis. The "U" shaped

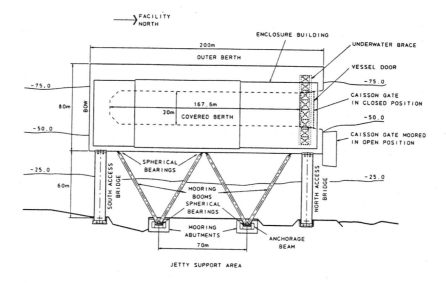

Fig. 1. Plan

post-tensioned concrete hull is 200m x 80m overall and provides a covered berth 167m long x 30m wide. Twin cellular pontoons, 25m wide x 12m nominal depth, are connected at the closed end by a rigid link and by a tubular steel underwater brace across the berth entrance. It floats with a draught of about 8.4m.

7. The enclosure building is formed by a steel framed superstructure clad with profiled aluminium sheeting.

8. The EHJ is moored on station by four tubular steel booms arranged in two V's. At the shore abutments, large rubber shear units provide the mooring flexibility.

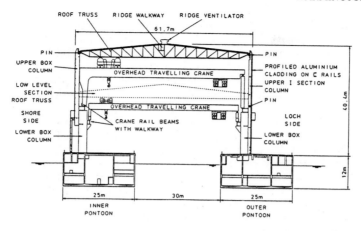

Fig. 2. Typical section

9. Fig. 3 shows there was considerable change in EHJ detail between the concept on appointment of the Consultants and the final design.

DESIGN

10. The EHJ has a briefed design life of 50 years with an allowance for two major refits. To satisfy the Brief, loadings with a return period of twice the design life, (i.e. 1 in 100 years), were used. In addition, for the Safety Case, the EHJ was checked for the consequences of 1 in 10,000 year events, (a particularly onerous design requirement), local damage being acceptable but retention of structural integrity had to be demonstrated.

11. The principal design loads are:-
- Dead Load
- Imposed loads, (from plant and machinery etc)
- Environmental loads, (wind, wave and current), and
- Hydrostatic loads

12. For the Safety Case, the following were also considered:-
- Earthquake
- Impact, (ship and vehicle collision and dropped loads)
- Flooding,
- Extremes of temperature and
- Systems failure

13. In addition to the normal checking procedures, the design was scrutinised by:-
- Independent audit teams within the Consultants' organisations.
- Design Review Committees both for the EHJ as a whole and for major individual components.
- Lloyds Register for certification.
- National Nuclear Corporation, (NNC), and the Safety and Reliability Directorate, (SRD), based on detailed input, prepared by the Consultants to the nuclear safety case documentation.

14. Both design and construction were also subject to a rigorous Quality

Assurance regime monitored internally and by PSA appointed auditors.

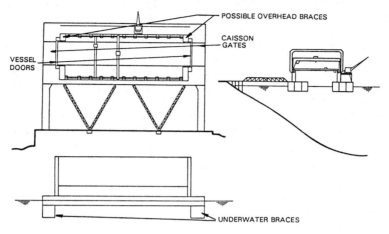

Fig. 3A. 1984

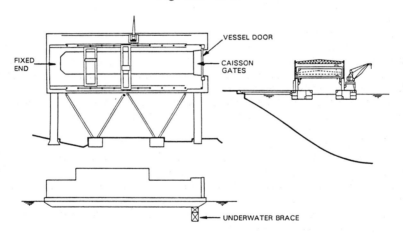

Fig. 3B. 1986
Fig. 3. Concept development

15. For wind loads consideration was given to the latest research into the UK wind climate. This showed significant reductions over British Standard CP3 ChV, but was not acceptable to the nuclear safety authorities. CP3 ChV was therefore used.

16. The mooring forces and EHJ and vessel response arising from wave and current were extensively investigated by Hydraulics Research Ltd, (HR), in large scale physical model studies. HR established the wave climate using their HINDWAVE program with spreading functions appropriate for the loch. Both long and short crested random waves were modelled, calibrated to a JONSWAP spectrum. The latter, which are more representative of a true sea state, produced significant reductions in

mooring forces over the former, and were adopted in the final design.

17. Operational response studies were also undertaken by HR, which included the use of a wind tunnel in conjunction with the wave tank.

18. Initial work by HR showed that a flexible mooring system greatly reduced mooring forces and this was therefore adopted. Whilst this also had the advantage of filtering out response to earthquakes, work by Dr.T.Wyatt of Imperial College showed that, despite the EHJ's size, the flexible mooring system was potentially susceptible to wind induced dynamic response. A detailed assessment of this was therefore undertaken.

19. Peak wind induced mooring forces were calculated from statistical data on total EHJ wind loading, (established by Oxford University using a static wind tunnel model), modified by spectral analysis techniques to account for dynamic response of the mooring system. Total mooring forces were established by statistically combining these with the results of HR's work.

20. The hull design was undertaken combining the output from three separate analyses:-
- Finite Element analysis to establish global stresses for static and quasi-static, (wind, wave and current), loading.
- Finite Element analysis to establish global stresses for earthquake loading, and
- Local panel analysis.

21. The large static FE model used membrane elements to represent the hull and beam elements to represent the mooring booms and underwater brace. The superstructure was only partially represented to obtain deflections at critical positions. Springs modelled buoyancy and the flexible mooring. Load combinations were applied to the model to produce the most onerous conditions of hull hogging, sagging, torsion and differential pontoon flexure.

22. For the earthquake analysis a simpler beam model of the hull was used with a more detailed representation of the vessel access door, (VAD), support structure. A comparison of the primary global motion frequencies obtained from the FE analysis and experimentally by HR showed excellent correlation. Near and far field earthquakes were considered, with a zero period acceleration of 0.2g horizontally and 0.135g vertically. Also two types of analysis were undertaken:-
- Response spectrum for checking the hull, underwater brace and mooring booms, and
- Time history to assess the effect on the superstructure, cranes and VAD.

23. Because of the filtering effects of the flexible mooring and floatation, earthquake was not a critical design case, even with a 40% "cliff edge" increase to test the design sensitivity.

24. Extensive post-processing of the static and dynamic FE models was undertaken using in-house developed programs. This included computerised stress and fatigue checking of the underwater brace to API codes and the production of earthquake frequency/acceleration spectra at various points for checking the superstructure, cranes and VAD.

25. The superstructure, mooring system and caisson gate etc were designed independently from the hull, due allowance being taken of the output from the hull analysis as appropriate.

26. Fig 4 shows a typical cross-section of the concrete hull. A 50N/mm² concrete was used with a specification aimed at achieving high

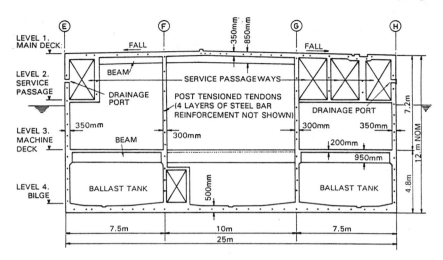

Fig. 4. Typical hull section

durability and water-tightness. As the hull is some 88% of the total EHJ weight, wall and deck thicknesses were made the minimum practicable for construction, (consistent with avoiding shear reinforcement), and to accommodate the ever increasing M&E spatial requirements. High yield reinforcement was used, quantities generally being governed by crack width considerations. Cover was varied from 30mm internally in dry cells to 75mm externally in the splash zone. Post tensioning was provided longitudinally and across the link end to cater for longitudinal bending and torsion, with some 258 tendons up to 190m long. These were grouted after stressing to provide corrosion protection and to prevent explosive energy release in the event of collision damage.

27. The EHJ design Brief contained detailed criteria for the geometry including:-
- Plan area layout
- Pontoon construction depth
- Pontoon cross-section and
- Operational requirements, including a 5m freeboard.

28. As a consequence of design development and, in particular, the increasing weight of the three major components, (hull, superstructure and M&E), it became apparent that all these requirements were not mutually compatible. Given the retention of the overall dimensions, which produced a highly efficient design, both structurally and in respect of total weight, the actual freeboard was reduced to 3.6m.

29. The underwater brace is a tubular steel braced frame structure connecting the arms of the EHJ at the berth entrance. It is attached to each pontoon by four legs which were stressed and grouted into shafts formed in the hull structure. The brace was designed to provide a similar stiffness

to the concrete link at the closed end of the EHJ. Also, in the event of damage, it has sufficient reserves of strength to function with only one half intact.

30. Fig 2 gives a typical section through the superstructure. It is a steel frame with profiled aluminium sheet cladding.

31. Fig 5 shows a typical superstructure frame. The columns are

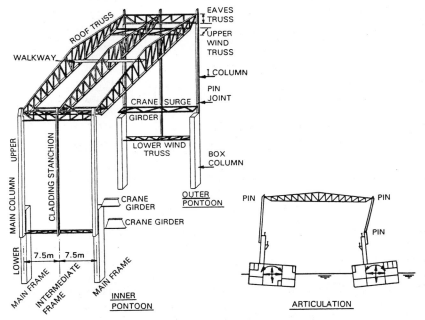

Fig. 5. Typical superstructure frame

generally at 15m centres with roof trusses at 7.5m centres. Pin joints permit the structure to accommodate hull flexure. As a consequence the majority of transverse load is carried by the inner pontoon columns which are therefore the more substantial. The columns are stressed to the hull by up to 36 Macalloy bars. Temporary sliding bases were provided to certain columns to allow for hull shortening during post-tensioning.

32. The VAD support structure is a free standing braced frame portal spanning between the two arms of the EHJ. It contains all the VAD machinery and has a non-structural suspended facia behind which the doors retract.

33. Movement joints were provided in the superstructure at the changes in building height, at the centre of the high level section and at the VAD portal to accommodate hull and superstructure flexure. The crane rails however are continuous over the expansion joints.

34. The mooring booms are 2.5m diameter stiffened tubes of varying wall thickness. Splice joints were introduced into two of the booms to allow for construction tolerances. These consist of a shimmed joint stressed together with Macalloy bars around the outer circumference.

Large plain spherical bearings are at each end of the mooring booms, stressed to the shore beam, booms and hull by Macalloy bars. The booms and the EHJ end bearings were designed for working loads of 985 tonnes and extreme loads of 1800 tonnes.

35. On shore, 24 large rubber shear units are symmetrically placed above and below a fabricated steel anchorage beam at each mooring abutment, (Fig 6). These provide the required mooring system flexibility

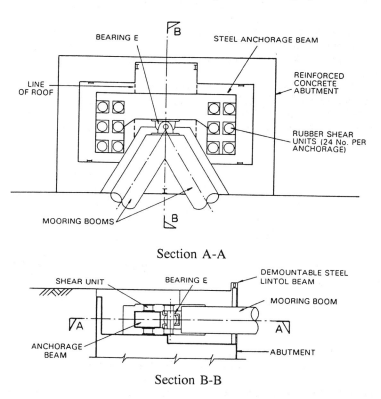

Fig. 6. Mooring abutment

and were designed for working loads of 1315 tonnes and extreme loads of 2280 tonnes. The reinforced concrete abutments are secured back to the shore with arrays of rock anchors.

36. The shear units were moulded from a modified natural rubber formulated to meet the design requirements. The design of the units by a specialist sub-contractor involved an extensive programme of theoretical design, include fatigue analysis using fracture mechanics, material and model testing and full scale testing, including fatigue simulation. From an early stage of the design it became apparent that, unlike the rest of the EHJ, fatigue was a primary design consideration. In order to improve fatigue resistance seven bonded steel plates were introduced into the units. This had the consequence of greatly increasing the axial stiffness of the

units and necessitated strengthening of the anchorage beam and abutment roof to cater for the greatly increased vertical tensions induced in the shear units at high shear deflections.

37. As the shear unit design developed a conflict between the requirements of the specified linear load/deflection characteristics, ultimate strength and fatigue strength was resolved by increasing the number of shear units at each abutment from 20 to 24, relaxing the shear stiffness linearity requirements and assessing the effects of this on the mooring loads.

38. The successful culmination of this work was the testing of a full scale unit in a simulated fatigue test followed by an overload on the extreme load of 34% with no signs of incipient failure or permanent damage.

39. The access bridges are trapezoidal box girders of 61.27m span. Unusual aspects of the design were the need to consider storm wave loading and the requirement of the bearings to accommodate large plan movements generated by EHJ motions, (Fig 7). The bearings at the

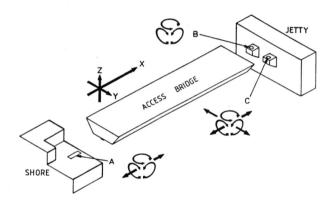

Fig. 7. Access bridge bearing motions

pontoon end, (B and C), are plain spherical bearings. Bearing B pins the bridges to the EHJ and is free to rotate in all three axes. Bearing C is free to rotate in all three axes and translate along and, (to a lesser extent), perpendicular to the bridge axis; achieved by a linked double bearing. Stability to the link mechanism in bearing C is provided by bearings A and B through the bridge.

40. Bearing A, at the shore end, has to cater for large movements along the bridge axis, (up to ±1.43m in extreme storms), as well as rotation in the three planes. A standard bridge bearing, with an extended sliding plate to accommodate the linear motions, was initially envisaged. However, in the event this was ruled out due to problems of wear on the PTFE bearing surfaces arising from the fretting motions to which the EHJ subjects the bearing. The solution is a system of bearings for the rotations mounted on a bogie running on rails. Both the rails and bogie wheels are face hardened with Stellite to counter fretting wear.

41. Allowance for removal of Bearing A for major overhaul required the provision of a jack mounted on Hilman rollers running on rails adjacent to the bearing. The design of this system allows for EHJ motions in the prescribed maintenance weather window.

42. The caisson gate is a steel structure. When in place it is ballasted onto bearings on each of the EHJ arms to give some 300 tonnes preponderance. It is stepped and unstepped by pumping water to and from ballast tanks and is winched into and out of position. As well as closing the berth, it also acts as a bridge between the two arms of the EHJ.

43. All steelwork has galvanising and/or paint systems for corrosion protection. For steelwork in contact with water, sacrificial zinc anodes are also provided. Bridge and caisson gate decks have an epoxy based wearing surface with aggregate dressing. The aluminum cladding has a bonded colour coat chosen to allow, as far as possible, the EHJ to complement its surroundings.

44. At the heart of the Facility are two unique high integrity electric overhead travelling cranes, designed and built to meet stringent nuclear safety requirements. Both cranes, as far as practical, are of dual load path design, allowing mechanical and/or electric components to fail without resulting in a dropped load. Duplicate drive path features enable loading operations to be completed in the event of equipment failure. Of particular note are the crane hooks which were developed by analysis and extensive testing to meet the stringent service requirements. A special steel composite was developed with a particular heat treatment specification to give a high strength material with good fatigue properties. The cranes were subject to seismic, fatigue and failure mode and effect analyses.

45. The VAD comprises a two leaf door, providing a clear opening 37m wide by 15m high above the jetty cope. The door leaves, which each weigh about 65 tonnes are fully balanced by counterweights and operated by separate winch systems.

46. Both the crane and VAD designs cater for hull motions and flexure. This was achieved on the cranes by the outer pontoon crane legs being pinned at their junction with the cross beams. The VAD has sprung wheels running on the vertical guide rails.

47. Critical nuclear related power supplies within the EHJ have been duplicated and physically segregated to maintain nuclear integrity through defined hazards. In addition, a diesel generator has been provided as a standby for EHJ operational loads, (excluding the vessel), in the event of shore supply failure. A frequency changer in each power train converts the 50 Hz supply to the 60 Hz requirement of naval vessels.

48. Extensive pipework distribution and pumping systems were installed to service the EHJ and vessel. These include fresh water, salt water fire fighting, sewage storage, compressed air, demineralised water plant and bilge and ballast systems. Pipework is generally Cupro-nickel to give a 17 year design life in a marine environment. Expansion bellows and articulated joints were incorporated to allow for hull shortening during the stressing operation, relative movements at mooring booms, (which serve as pipe bridges from the shore to the EHJ), and thermal expansion due to extremes of temperature.

49. Pumped bilge and ballast systems maintain the integrity of the EHJ

under operational and emergency situations. The systems were duplicated and designed to meet extreme events, including flooding resulting from vessel collision and burst pipework.

50. Computerised System Control And Data Acquisition, (SCADA), systems and computer controls have been provided to allow monitoring and remote operation of engineering systems from locations both on the EHJ and shore.

51. The whole electrical installation was designed, installed and tested against an extensive Electro-Magnetic Compatibility, control plan and the M&E systems were tested and commissioned to the stringent requirements of the Nuclear Safety Case.

THE CONTRACT

52. A target date of mid 1987 was set for the issue of tender documents and, with this in mind, a prequalification exercise was commenced. Major Civil Engineering Contractors were approached to determine their interest, capabilities, financial structure and proposed construction sites. The latter item featured very strongly in the formulation of a tender list.

53. At this stage there were a variety of construction options;
- to construct totally in the dry, launch and tow to Coulport.
- to construct the hull in the dry and fit-out alongside a berth
- to construct the hull in sections and join afloat.

54. A number of potential sites were visited to determine their suitability and availability with particular regard to:-
- distance from Coulport
- security arrangements
- deep sheltered water
- site services and buildings
- access and
- availability of labour

55. As a consequence of these visits and consideration by the PSA and the Consultants of technical and financial matters, a tender list of 6 contractors was produced and conventional bid documents issued. Tenders were returned indicating 6 competitive prices for a concrete hull with only 2 for a steel hull; the latter were sufficiently high to preclude them from further consideration.

56. Following scrutiny of tenders, including method statements, qualifications, prices and programme, a contract was awarded by the PSA to a joint venture of Costain - Taylor Woodrow, (CTW), in October 1988.

CONSTRUCTION

57. The tender of CTW for a concrete hull construction was based on the use of the former ANDOC yard dry dock, situated in the Firth of Clyde between Hunterston Ore Jetty and Hunterston Power Stations. This was constructed in the early seventies for offshore oil rigs and is a deep basin formed within a man-made dredged sand island.

58. This yard had a number of significant advantages for CTW, namely:-
- good communications, being adjacent to the A78 coastal road.
- access to a local workforce

- good existing buildings including a 2 storey office block and 2 large storage sheds, each with overhead cranes
- reasonable security; the "island" being linked to the mainland by a 2 lane causeway
- existing services in place - power, water, telephones and a sewage treatment system
- an existing and operating well point drainage system
- deep water channel adjacent to the dock
- reasonably sheltered waters for the tow to Coulport

59. CTW commenced by laying down a bed of 1.6mm steel plates on no-fines concrete over the entire base area, (10,300 sq m), of the EHJ to form the soffit of the bilge slab. The no-fines concrete allowed the free passage of water under the EHJ to facilitate float-up during dock flooding. The basin was also extended to accommodate the full length of the hull.

60. Two site batching plants were set up, each having a capacity of 40 cu m/hr. High quality coarse aggregate was obtained from Hillhouse Quarries in southern Ayrshire and crushed sand from Douglas Muir. A low alkali cement from Germany was supplied by Castle Cement with 30% PFA obtained from UK power stations.

61. Construction of the hull commenced at the closed end and progressed by stages along each arm towards the open end. The design, as described, required relatively thin walls to be built in a series of contiguous cells. A series of tower cranes were employed to handle materials, and concrete pumps were extensively utilised. Given the high density of rebar, the added complexity of post-tensioning ducts and their stressing, the need for watertight construction joints and the difficulties of general access for working within confined cells, CTW are to be congratulated on achieving a high standard of concrete work.

62. The next major stage of construction related to erection of the steel superstructure. CTW's sub-contractor for this work was Victor Buyck (UK) who fabricated the steelwork in their Belgian shops and shipped it to Hunterston by road and sea. Concurrently CTW's marine steelwork sub-contractor, (Ravestein BV), were fabricating the mooring booms, box girder bridges, underwater brace and caisson gate in Holland.

63. Most of the latter steelwork was shipped by barge and stored at Hunterston, the exception being the 650 tonne underwater brace which was shipped in one piece aboard a special vessel, the Fairmast Jumbo. The Fairmast anchored off Largs and lowered the brace into the sea using its two 500 tonne cranes, whence it was towed, floating on its side, to a temporary anchorage off Great Cumbrae Island.

64. Throughout these intense activities, work continued on fitting out the EHJ with M&E plant and equipment including the diesel generator, frequency changers, various fluid tanks, the 2 cranes, (weighing a combined 1360 tonnes), and the VAD. The construction logistics were dominated during this stage of the project by the parallel activity of post-tensioning of the hull.

65. A temporary brace was required, inboard of the permanent underwater brace position, to maintain structural integrity of the hull during tow and until installation of the permanent brace at Coulport. The draught of the permanent brace requires about 26m of water which precluded fitting

before the tow.

66. Finally in March 1992, the EHJ was ready for dock flooding. This was accomplished by pumping continuously over an 80 hour period using a cutter suction dredger operated by sub-contractor Van Oord ACZ. The dock bottom was at a level of -12.4m O.D, thus providing sufficient water depth for the EHJ to float at all states of the tide after breaching of the outer bund by the dredger. At this time the EHJ was tethered with 8 heavy mooring ropes to strong points on the dock sides. At the final predicted draught of 8.18m, EHJ became a floating structure. To the credit of all parties the subsequent inspection showed no leakage through the hull. In particular the steel bed plates had served their purpose, peeling cleanly away from the bilge to leave a surface requiring no remedial works.

67. The marine steelwork had by now been shipped to Coulport. The booms and bridges were swung into position on their shore bearings, supported at their offshore ends by a pair of north-sea type barges.

68. A suitable weather window became available on the 6th April 1992 for the tow to Coulport. Meticulous planning and attention to detail by CTW and their marine sub-contractor, Neptun, resulted in an outstanding success for the 38 km tow. On a rising morning tide, the moorings were successively cut, followed by a progression of tugs into the dock to maintain position and control of the EHJ. At 14.00 hours the EHJ began its 10 hour journey to Coulport, fixed end to the fore, headed by 3 tugs with 2 further tugs astern providing braking.

69. The tow, on a fine afternoon, was watched by a large number of interested spectators as the procession moved up the Firth of Clyde at a stately 3 knots. The EHJ arrived, without incident at Coulport and over the next few days was permanently attached to the mooring booms and the bridges were located. The standard of workmanship had been such that the fit of the booms and bridges were well within the allowable tolerances.

70. One of the most difficult operations accomplished by CTW/Neptun was the installation of the permanent underwater brace. This was towed to Coulport lying on its side, virtually submerged. Installation involved ballasting the brace into a vertical position, 2 of its 8 vertical legs being topped by stabbing guides of different lengths. A special barge then straddled the brace which was further ballasted and lowered beneath the EHJ keel and the barge brought into the inner berth. The legs were positioned under the circular shafts in the pontoons and winched up until collars on the legs made contact with the bilge slabs. Following the dewatering of each shaft and the stressing and grouting of the underwater brace legs to the hull, the temporary brace was dismantled and removed.

71. The remainder of the work was largely routine, consisting principally of the completion and setting to work of the various M&E systems, inclining tests on the caisson gate, and identification and rectification of defects and snags. Formal handover to the MoD was marked by a ceremony on the EHJ on the 23rd March 1993 conducted by Vice Admiral Sir Neville Purvis KCB.

72. Fig 8 shows the EHJ on station at Coulport.

73. At the peak of construction activities a total of 200 staff and 900 labour were employed. Industrial relations were generally very good, and the CTW safety record was exemplary; a few minor accidents but no

fatalities occurred.

Fig. .8. The EHJ on station

74. The principal CTW sub-contractors who contributed to the success of the construction phase were:-

- Babcock Energy Ltd   -   Vessel Access Doors
- Bakker Rubber   -   Mooring Shear Units
- Balvac Whitley Moran   -   Pontoon stressing
- Ives Cladding   .- Superstructure Cladding
- James Scott   -   M&E Services
- NEI Clarke Chapman
  (Wellman Booth)   -   Cranes & Bridge Bearings A.
- Neptun   -   Tow and major Marine Works
- Ravestein BV   -   Marine Steelwork
- SKF (UK) Ltd   -   Bridge and Boom Bearings
- Van Oord ACZ   -   Dredging and Bund Reinstatement
- Victor Buyck (UK)   -   Structural Steelwork

CONCLUSION

75. The EHJ is testimony to the dedication of many professionals in diverse disciplines, producing a unique and complex floating structure; from original concept through many years of development to its present status as a major component of the Clyde Submarine Base.

76. The EHJ is most certainly unique and complex. The overall spatial parameters specified to the designers resulted in considerable difficulties in accommodating a variety of M&E systems and equipment including 200kms of cables and 12kms of pipework.

77. There was considerable growth in the M&E component of the works during the project. The ratio of Building & Civil Engineering to M&E costs has reversed from 2.8:1 at concept, through 1.3:1 at tender to 1:2 on completion.

78. Some doubts have been expressed over the years on the wisdom of selecting concrete as the construction medium for the EHJ. However the choice of concrete has been vindicated by a watertight and highly stable structure which is a tribute to all concerned.

Design and construction of the jetty access roads and support area

B. T. THOMAS, Projects Group Manager, Acer Consultants Ltd, and
E. R. SHARPLES, Contract Director, Tarmac Construction Ltd

SYNOPSIS. JARSA Contract involves the construction of a Jetty Support Area (JSA) and an extensive network of access roads. The JSA is bounded on one side by a 480 metre long sea wall which incorporates the anchorages and bridge abutments for the floating Explosives Handling Jetty and on the other by steep excavated rock faces up to 45 metres high. 6 kilometres of access roads connect the new jetty with the existing RNAD Coulport and the new Trident storage and maintenance facilities. This paper outlines the design and construction processes with particular emphasis on user requirements and the topographical, logistical and environmental constraints of the site.

INTRODUCTION

1. Acer were commissioned in May 1984 through the Property Services Agency as Lead consultants on the design and construction of a major construction package forming part of the Extension to RNAD Coulport. The package comprised approximately 6km of 7.3m wide road and a Shoreside Support Area in support of an 85,000 tonne floating jetty to be used for weapons handling. The final layout of the works is detailed in Figure 1. As Lead Consultants, Acer were responsible for coordination of a multi-disciplinary design team, comprising:

Acer Consultants	Lead consultant/Highway & Structural Engineering
Building Design Partnership	Architects
Entec (Wallace Whittle & Ptnrs)	Building Services
Kennedy & Donkin	Services distribution & Fence Security System
James Gentles & Son	Quantity Surveyor
R P S Cairns	Landscape Consultants

2. Prior to the commencement of the detailed design, sketch plans were produced by the Design Team and presented to the PSA and MOD to ensure that the users requirement had been properly interpreted. Detailed design commenced in early 1986 incorporating PSA/MOD comments.

3. Tender documents were issued in March 1987 and a contract awarded to the successful tenderer, Tarmac Construction Ltd - Major Projects Division in

October 1987. Work commenced on site in November 1987 and, following a phased handover, were completed in August 1992. The principal sub-contractors involved in construction of the works were as follows:

Ritchie/Rocklift Joint Venture	Drill and Blast
	Rock Face Stabilisation
	Ground Anchors
Archbell Greenwood	Structural Steelwork
Andrew Young & Son	Structural Steelwork
MCL	Cladding
Matthew Hall	Mechanical/Electrical
Wimpey Asphalt	Coated stone
Shorrock Systems	CCTV & IDS
Avondale	Security Fences
Melvin Brothers	Safety Fences
Shiers/Shephard Hill	Diving Services
Hayden Maintenance	Security Gates/Turnstiles

The principal suppliers were:

Foster Yeoman (Glensanda)	Dry Stone
Topmix	Ready Mix concrete
Mannesman Demag	Gantry Cranes
Elcosta	Security Gates/Turnstiles
ASSET International	Safety Fences
Aberdeen Concrete	Precast Concrete Products
Hoogovens	Cladding
Eternit Cladding System	Cladding

LOCATION, TOPOGRAPHY & SOILS

1. Located on the Rosneath Peninsula, RNAD Coulport is part of the Clyde Submarine Base. Its function is the storage, preparation and servicing of both Polaris missiles and conventional torpedoes.

2. Coulport lies within an area of transition between the Highlands and the Lowlands. The Rosneath Peninsula is a glaciated ridge defined by Loch Long and Gare Loch; both major lochs forming part of the Firth of Clyde. At Coulport the ridge of the peninsula is defined less clearly with the topography typified by undulating ground with many knolls and depressions. A secondary plateau is evident above RNAD Coulport with steep slopes down to the boundary of the existing depot and, at its northern end, the shoreside of Loch Long. Over the area of the site rock outcrops at the surface it is covered with generally shallow layers of glacial till and humus.

SAFETY AND AVAILABILITY

1. JARSA was identified during its design phase as forming part of the RNAD Coulport Nuclear Safety and Availability Case and as such certain

elements have been designed to resist extreme environmental conditions including seismic effects. These elements are associated with the integrity of the EHJ including the supply of services within it. In this latter respect documentation produced in support of the JARSA Safety case has addressed the justification of service trenches, rockfaces and associated structures.

JETTY ACCESS ROADS
General Design Criteria

1. The Jetty Access Road Network links the Trident Storage Area with both facilities in the existing Depot and the Explosives Handling Jetty. The works involved approximately 6km of 7.3m wide single carriageway road with associated structures and service installations. The roads have a design life of 30 years with gradients, vertical and horizontal alignment all being limited by an overall design speed of 32km/hr.

2. All roads are kerbed and a positive drainage system provided by way of gullies and piped carrier drains. Surface water run off is picked up in interceptor ditches and channelled into the primary water courses. Similarly a trapezoidal channel located at the back of the verge picks up run off from the rock faces and carries it to the nearest water course.

3. The road pavement design was carried out using PSA Technical Instruction TICE 50 based on projected MOD usage. Where existing carriageway was to be overlain and reprofiled, a load survey was carried out using a Bankleman Beam to establish its condition and remaining design life.

4. On roads used by explosive transport vehicles a high containment safety barrier is employed. This comprises a double height, double sided open box beam safety fence using standard components. Standard barriers using an untensioned corrugated beam on wooden posts are used at other locations based on road usage.

5. Vehicle arrestor beds have been strategically positioned on long gradients for use in emergency situations. The beds were designed in accordance with TRL recommendations and comprise concrete troughs filled to road level with lightweight aggregate.

6. Numerous services diversions to existing RNAD Coulport and statutory undertaker services, both mechanical and electrical, were required. New and diverted services are located within the roadside verges.

7. The alignment of the roads network mainly follows the alignment indicated at the Environmental Impact Assessment stage. The roads were generally produced by cutting into sidelong ground for most of their length, producing a variety of cross sections depending on the steepness of the existing ground.

8. Owing to the topography of the area, significant disturbance to ground cover and land form was inevitable. However, every effort was made to minimise environmental impact. Existing vegetation cover on adjacent land was protected and retained until land use management issues were finalised.

9. It was considered environmentally important to preserve as much of the natural tree and scrub growth as possible and this was an ever present

JETTY ACCESS ROADS AND SUPPORT AREA

consideration in the minds of the design team. Ultimately the preserved growth was supplemented by strategic planting of compatible species of trees and shrubs.

10. The Network comprises five specific sections. The initial link, the Jetty Service Road, was provided under the MOD Advanced Works Contract to ease the importation of plant and materials utilised in the construction of the main works. The remaining four sections comprised the following:

   a) Explosives Area Access Road
   b) Administration Area Access Road
   c) Jetty Access Road
   d) Jetty Link Road

JETTY SUPPORT AREA

1. The Jetty Support Area is located on the steeply sloping shoreside of Loch Long, the precise position being dictated by explosives safety distances from the EHJ to the centres of population at Ardentinny and Portencaple.

2. The final layout for the JSA was developed from a study carried out in May 1985 which addressed the individual elements in 10 layout options. Figure 2 shows an aerial view of the completed works from the north.

3. The JSA is partially excavated from the steeply sloping hillside, an operation which resulted in the removal of approximately 100,000m^3 of rock in the formation of rock faces up to 45m in height. On the seaward side the JSA is bounded by a 480m long foreshore structure comprising a retaining wall, boom anchorages and bridge abutments associated with the EHJ. Fig 3 is a cross section through the JSA showing both the foreshore structure and rock cutting. Located within the JSA are installations required to service the EHJ:
   Support buildings
   Distribution points for incoming services
   Local services network
   Roads and hardstandings
The JSA is contained within a high security fence complete with closed circuit television and intruder detection systems.

FORESHORE STRUCTURES - DESIGN

1. The retaining wall structure is a gravity retaining wall and comprises precast interlocking concrete units, mass concrete, reinforced concrete and granular fill. It is founded on a bench cut into the bedrock of the steeply sloping hillside. To facilitate construction, extensive use is made of precast concrete (see Fig 4) which, when assembled behave monolithically. The detailed construction of the wall varies according to its height but Fig 5 and 6 show typical cross sections. The complete structure relies on its own weight to provide overall resistance to overturning and sliding. Principal loads are resisted by in- plane bending, out of place shear and static equilibrium. Tidal lag within the granular fill is neglected due to the presence of adequate drainage paths within the structure. The structure is inherently rigid and

effectively tied in the rock. Earth pressures are evaluated using 'earth pressures at rest' as the wall will not yield sufficiently to develop 'active' pressures. Water pressures are evaluated using conventional hydrostatic principles. A section of the retaining wall to the south of the southern bridge abutment is subject to the more onerous requirements of the Safety Case. The resulting structure (see Fig 7) incorporates sections of mass concrete and reinforced concrete counterfort wall all socketed into the bedrock and capped with a reinforced concreted tie back slab. The substrata beneath the founding level of the wall were pressure grouted using a cementitious grout. In this case the general design philosophy outlined above remains uncharged except that the Monocobe- Okabe procedure was used to establish seismic earth pressures and the Westergaard theory for establishing water pressures for seismic loads. Excess pore pressure in the rock is relieved by weepholes through the structure. However, due to the massive nature of the structure and the grouting operations to the bedrock, the effects of 'tidal lag' have also been considered.

2. The two access bridge abutments comprise the substructure surmounted by a reinforced concrete slab on which are located the bridge bearings. A curtain wall retains the fill behind the abutment. The bridge deck expansion joint is supported on a run on slab spanning back from the curtain wall and supported on counterfort walls. The substructure construction is similar to that employed for the retaining wall comprising hollow interlocking precast units and sections of mass and reinforced concrete with zones of reinforcement.

3. The two mooring boom anchorages maintain the EHJ on station at all times and limit the relative movements between it and the shore such that the access bridges do not become dislocated. The anchorages are essentially gravity structures, although each anchorage incorporates twenty four multi-strand restressable ground anchors to provide sufficient restraint to resist loads from the EHJ booms. Fig 8 shows the anchorage layout. All anchors have been installed and stressed in accordance with BSIDD81:1982. subsequently revised and reissued as BS8081:1989. Their working load is 1300k$_N$. All anchorages have a Vibrating Wire Load Cell and their load/temperature can be read through 4 No. Terminal Units with the use of a portable read out unit.

FORESHORE STRUCTURES - CONSTRUCTION

1. Construction of the foreshore wall presented a large problem regarding plant and method selection. The wall is founded at depth of up to 9m below the waters of Loch Long. Tidal working was employed where possible to minimise the amount of underwater working although there was still a requirement for some 50 divers at the peak of operations. The foreshore dips a 1 in 1 and consists of essentially a scree of weathered rock overlying hard schist bedrock. A temporary sheet piled wall for dewatering was therefore out of the question and even a causeway staging on steel H piles was discounted on the basis of the legs being difficult to found in the sloping rockhead and always susceptible to damage from the adjacent earthworks. Similarly, a large jack up barge was discounted because of the foreshore configuration.

2. The adopted method was a combination of land and marine based plant.

The drilling and blasting was carried out from a pontoon stationed vertically above the line of the wall after removal of overburden by land based backactors or drag line cranes. Working to instructed levels, underwater blasting techniques were used to fracture the trench for the block wall. The excavation was then carried out by modified long arm excavators breakers and grabs operating, for stability, from land. They were aided by pontoon based breakers able to operate against the tension of land lines and with the benefit of the self weight of the pontoon to which they were welded.

3. Initial setting out through to fine trimming was diver controlled with the assistance of various templates and scaffold frames designed to orientate the divers in near zero visibility whilst assuring the necessary clearance for the blocks.

4. The tasks was frequently iterative with bulk excavation quickly giving way to local trimming and removal of fines by air lifts. Most bays were finished off by divers using hand tools.

5. Once the excavation had been cleared, inspected for fractures and set out, the T shaped precast base blocks were hung from an adjustable lifting frame to a tolerance of 10mm on level, 10mm in alignment and 5mm between adjacent interlocking blocks. The blocks were then concreted either singly or in threes depending on the stepped formation level.

6. The cycle of setting and concreting had to be quick to avoid the silt which arrived on each tide. Over 3600 interlocking T shaped precast blocks were required to construct the wall from - 9m to +6.8m in 500,, lifts. The average block was 2.5m x 2.5m x 0.500m in size and 4 tonnes by weight. However, the design of the various structures within the line of the seawall resulted in the different types of block up to 7 tonnes and including long 40mm starter bars for the reinforced anchorages.

7. For the standard wall, once the base blocks had been set, subsequent layers were craned in relatively quickly and set on mortar beds by divers. these blocks contained hollow cores which were either filled with mass concrete or stitched with prefabricated reinforcement cages. all underwater concrete consisted an additive that negated the need for elaborate tremie systems and was placed with long boom mobile concrete pumps.

8. The wall alignment incorporated 20 changes of direction, many of which were achieved by diver placed steel shuttering.

BUILDINGS

1. A total of 7 No buildings are provided in support of the EHJ. These are referenced JSAB 1- 7 on the JSA layout plan (Fig 2). Five of the seven buildings are of RC construction and brickfaced with roofs of profiled aluminium sheeting on a steel frame.

2. The remaining two buildings are of industrial type construction, steel framed with frangible cladding. The larger of these two buildings contained a 30 tonne overhead travelling crane.

3. The design strategy relating to buildings concentrated on establishing principles based on form massing, siting and colour. This created a low key

but appropriate family of buildings avoiding powerful three dimensional forms. The colours of materials were chosen to blend as much as possible into the existing landscape.

Building functions include:
- Access control
- Sub stations
- Storage
- Weapon operations
- Personnel occupation

Buildings in the JSA have been designed to criteria consistent with their location and function. These include blast overpressure and security hardening.

EARTHWORKS - GENERAL

1. Earthworks for the access roads and support area involved the excavation of some 144,500m^3 of superficial deposits and 434,300m^3 of rock, distributed as follows:

Location	Superficial m^3	Rock m^3
Explosives Area Access Road	41,400	73,400
Administration Area Access Road	43,300	100,700
Jetty Access Road	8,600	25,500
Jetty Link Road	32,400	109,900
Jetty Support Area	18,800	115,800
	144,500	434,300

2. The topography of the site and geometric design criteria for the road meant that there was a considerable surplus of excavated material to be disposed of. Results from a Development wide study pointed to limited infilling of the foreshore to the south of RNAD Coulport as being the preferred option for the disposal of the majority of the surplus excavated material on the JARSA contract. The remainder was removed to other designated disposal areas local to the works.

3. The superficial deposits, generally less than 1m thick, comprise sandy soil with schist gravel, cobbles and with topsoil or peat.

4. Bedrock consists of interbanded pale grey quartz.schists, greenish grey mica- schists, both strong to very strong, and pale green with a waxy feel weak chlorite schists. White quartz, in the forms of veins and lenses occurs throughout the bedrock. The schistosity (foliation) is the most dominant structural feature in the rocks and its general orientation is consistent throughout the site with dips to the south- southeast inclined at between 50 degrees and 70 degrees. In addition to discontinuity associated with the schistosity, seven other joint sets have been identified from mapping. The spacing, continuity and orientation of these joints have been found to be highly

variable but, from the stability point of view, the most problematic joint set was the generally westerly dipping one. Faulting, which is generally consistent with the foliation, has been identified and comprise clay filled shears typically 100mm wide.

EARTHWORKS - SLOPE GEOMETRY

1. The stability of the rock cuttings is controlled by the discontinuities within the rock mass. In deriving the slope geometries, the possibility of minor faults dipping at 65 degrees or greater to the east or southeast, was recognised. It was also assumed that the westerly dipping joints could dip at any angle between 20 degrees and 90 degrees with dip direction ranging from 230 degrees to 360 degrees, based on the recorded observations. Bearing in mind the natural topography of the area and from the assessment of the stability checks, a maximum overall slope angle of the order of 55 degrees was adopted. This angle was produced by forming 10m high faces at 65 degrees to the horizontal with 2m wide intermediate berms. Rock faces were limited to a maximin on lift of 10m in order that the drilling accuracy required by the specified perimeter blasting techniques could be readily achievable. Where land take permitted, slopes in highly fractured rock were trimmed back to 50 degrees to the horizontal to limit their susceptibility to spalling and ravelling from frost action and weathering.

2. Slopes in superficial deposits and weathered rock were cut at 1 (vertical) to 2 (horizontal).

3. Generally the roads are formed in sidelong ground, part in cutting and the remainder on embankment. The embankments were up to 10m in height and benched into bedrock. These benches were formed with a maximum vertical height of 1m and drained towards the outer edge of the embankment. Embankments were constructed generally with side sloped of 1 (vertical) to 2 (horizontal) using rock excavated from the works.

EARTHWORKS - EXPLOSIVES AND BLASTING

1. It was evident in the early stages of the JARSA design phase that extensive rock blasting would be necessary if the stated programme objectives were to be achieved.

2. Discussions with PSA and MOD ensured that concerns regarding the effects of blasting on the day to day working of RNAD Coulport were identified. Trials were carried out pre contract to investigate bedrock propagation properties, from which Acer were able to advise on acceptable blast vibration levels at particular buildings within RNAD Coulport. In additional a full condition survey was carried prior to blasting operations commencing.

3. Based on the above discussions and trials, a specification was produced aimed at ensuing compliance with MOD's stringent requirements for explosives safety, operational needs and long term maintenance on excavated faces.

4. The specification required that the final faces of all rock slopes of 50

degrees or more to the horizontal be achieved by means of controlled perimeter blasting techniques. This included presplitting, smooth blasting and line drilling.

5. Following site trials the majority of presplitting was carried out using 100mm diameter holes spaced at 750mm centres, although this latter figure varied between 300mm and 1000mm as required by the condition of the rock. One of the factors governing the effectiveness of presplitting is the accuracy of the drill hole relative to the final face and, to this end, "down the hole" hammers were used for all controlled perimeter drilling. The setting up of a drill rig at each location was the subject of rigorous on site checking.

6. The philosophy of the Project Specification for blasting was one of failure to safety. Every stage of the blasting operation was under the control of the Tarmac's blasting engineer and work monitored by the site control team to ensure that the risk of misfire was minimised. Special provisions were made to ensure that multi- point initiation was used which would provide duplicate back up in the event of either detonator or primer failure.

7. The choice of initiation system was given the same level of consideration as the explosives themselves. Environmental and noise restrictions meant that surface lines of detonating cord could not be used although its use 'inhole' was permissible. The choice of Nobel Explosives 'Magna' detonators recognised that protection from radio/radar interference would be required to prevent inadvertent initiation by stray currents.

8. To avoid the risk of unexploded detonators remaining in the ground on completion of the works there was a requirement in the specification for all detonators to be located within the excavation envelope.

9. Modern slurry and emulsion explosives were use exclusively. These explosives give all the advantages of nitroglycerine based compounds but, in the event of a misfire going unnoticed, do no present the same risk, as the properties of slurries and emulsion ensure that degradation under normal atmospheric condition is rapid and results in the breakdown of the compositions into inert substances.

10. All blast proposals were vetted to ensure that the permitted levels of ground vibration contained in the contract were not exceeded. Further, all blasts were monitored for ground acceleration/acceleration limits and the results obtained compared with the theoretical value contained in the blast proposal.

11. Blasting was successfully employed to develop approximately 50,000m^2 of presplit face. In some instances blasting was carried out within 10m of existing buildings and throughout vibration levels were contained below the limit set in the Specification.

EARTHWORKS - SLOPE STABILISATION

1. Both tensioned and untensioned rockbolts were utilised as part of the stabilisation works. Tensioned rockbolts were used to secure blocks of rock that were either too difficult to remove by barring or so big that their removal would impair the stability of adjacent parts of the rock face. They have also

been used to restrain blocks which were defined on their base by relatively steep westerly dipping joints. Tensioned rockbolt lengths varied between 4m and 8m with a working load capability of between 150KN and 250KN. These bolts were primary and secondary grouted to provide full column bonding for longterm stress transfer considerations. Further, 'double corrosion' protection was provided to rockbolts consistent with the recommendations of BSI DD81:1982 subsequently issued as BS8081:1989. Untensioned epoxy coated, fully grouted rockbolts have been used where face support only was required. A total of 1544 tensioned and 4435 untensioned rockbolts were installed, of which 2086 of the latter were in the rock cuttings at Kibble Gardens.

2. Sprayed concrete of minimum thickness 150mm, reinforced with welded mesh and tied into sound rock using rockbolts was employed as a means of supporting faces containing loose/fragmented rock. Drainage paths were provided through the sprayed concrete by drilling weepholes at frequent intervals 1m - 2m into the rock mass. Approximately 26,600m^2 of the exposed rock faces were treated using sprayed concrete.

3. For local stability problems, ad hoc support systems were applied to suit the geometry of the face. Tied reinforced concrete walls were used to support individual blocks of rock liable to roll or topple and to underpin overhangs. Where exposed, weak and erodible zones such as faults or chloride rich horizons, were cut back and the resulting void infilled with concrete. Both insitu and sprayed concrete with tensioned or untensioned rockbolts have been used in the construction of tied walls. Weepholes were provided through the concrete to limit water pressure built-up behind the wall.

4. Drainage of the west dipping slopes was provided to minimise ground water pressures in the immediate vicinity of the faces and thus improve overall stability. Natural drainage paths exist through the persistent foliation joints, but these were supplemented by positive drainage measures particularly in those areas where rockbolts and shotcrete were installed. Drainage holes in the face were typically between 1m and 3m in length where face support was provided and between 5m and 30m in unsupported areas of the face. Slotted or perforated pipes were used to line the drainage holes where required by rock conditions.

5. Plastic coated chain-link mesh (to BS1722 Part 10) tied back to the rock mass using untensioned rockbolts was employed to restrain areas of loosely jointed rock prone to frost action and rainwash.

6. Rock face mapping was carried out after the bulk excavation to both record the geological structure and identify any areas of potential instability. Rock stabilisation details were superimposed onto drawings showing the rock face mapping to provide as installed drawings to assist in long term monitoring/maintenance of the cut face.

PROJECT MANAGEMENT

1. The JARSA package is complex in its own right and there were also major design/construction interfaces to be managed with form of other contracts of similar complexity:

Trident Explosives Area
Floating Jetty
Non- Explosives Area
External Service Distribution

2. A further major interface existed between JARSA and the RNAD Coulport, in particular the 1.2km of access road contained within the secure perimeter fence.

3. The scope of the Works involved nearly every conceivable construction task from the creation of working platforms through to M&E commissioning. This demanded a multi- discipline team and an organisation to cope with the combined logistical problems associated with the project.

a) Safety
Blasting in an explosives depot.
Multiple service diversions/permits to work.
Working over and under water.
Working on 40 metre high rock faces.
Parallel working in small areas.

b) Security
Working in and adjacent to a live Naval base with controls on personnel, shipping movements and working patterns.

c) Quality
Control of a diversity of tasks from pre- cut batters to special painting, a critical section of the Works being to seismic level of control.

d) Environmental Matters
The site is remote from centres of population but during the development period of the Clyde Bases the area became a hotpoint of construction activity. Movement of resources and materials was a major factor in the planning debate, with sea delivery of bulk materials for the Works a specified requirement.

e) Site Conditions
Rainfall 1800mm per annum, steep accesses, high winds and heavy seas possible in an exposed sea lock.

4. The method chosen to overcome the multiple logistical problems was to treat each one separately and attempt to strip away or minimise their effect on the people at the workface. The separate teams dedicated to each problem provided the possibility of single point contact with Contractor, Site Control Team, Client and Subcontractors alike.

5. A small team ran the importation of sand, gravel, P.F.A., cement, fill material, and roadstone for not only the JARSA contract but the other two

main contracts on the development. Shipping was limited to a stone capacity of about 2500T to match the dock design and the team worked a flexible rota to match the arrival of ships on the tides.

6. Similarly, to minimise the security restrictions a small team was made responsible for screening with M.O.D. many thousands of security applications for employees, subcontractors, suppliers, plant and vehicles. Recruitment was largely based on Glasgow and Paisley with transport arranged for day shifts, night shifts and tidal work.

7. On a larger scale the two main areas of the site were planned and resourced separately to allow each to follow its particular work patterns and planning priorities at operational level. These two areas were:

a) The existing RNAD Polaris Depot, where work was finely tuned around permits to work and a higher level of security, giving a short effective working day and smaller mobile plant.

b) The new development area where work could continue round the clock to follow the tides and weather conditions with only occasional security stoppages.

Each area was linked only by the minimum management necessary to ensure specialist subcontract and material control but benefited from the logistical cells organising security, quality, materials and the like.

8. A similar broad view was taken on safety that risks increase with parallel working in congested areas particularly at different levels. Where possible, tasks were scheduled to cut interfaces and to keep the arterial access routes open; for example, divers working during the day shift, selected excavation and restocking of materials at night. It was critical to keep material stocks around the construction zone as light as possible to reduce congestion and maintain safe access and egress. Similarly, during the Spring and Summer the divers and rock breakers were pontoon based to provide the maximum space for land based plant to manoeuvre.

9. The adopted Quality System drew upon Tarmac's earlier experience on C.S.B. Faslane and again required fundamental decisions to improve communications and control. Works procedures were produced from the bottom up rather than enforced downwards and were extended from the permanent work through to internal systems such as plant requisition and security clearance. This proved invaluable for the integration of new staff in a team which naturally evolved over the period of the Contract.

FIG 1 : JARSA - SCOPE OF WORKS

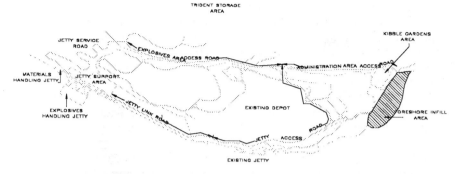

FIG 2

FIG 3 : TYPICAL CROSS-SECTION THROUGH JETTY SUPPORT AREA

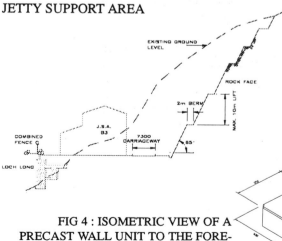

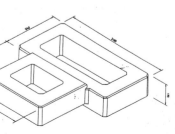

FIG 4 : ISOMETRIC VIEW OF A PRECAST WALL UNIT TO THE FORE-SHORE RETAINING WALL

JETTY ACCESS ROADS AND SUPPORT AREA

FIGS 5 & 6 : TYPICAL FORESHORE WALL CROSS SECTIONS

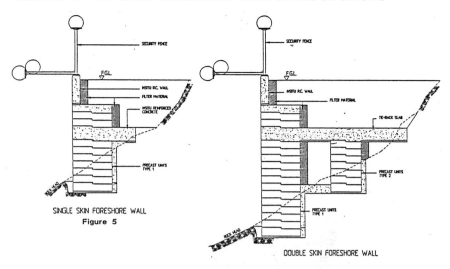

FIG 7 : FORESHORE RETAINING WALL SEISMIC SECTION

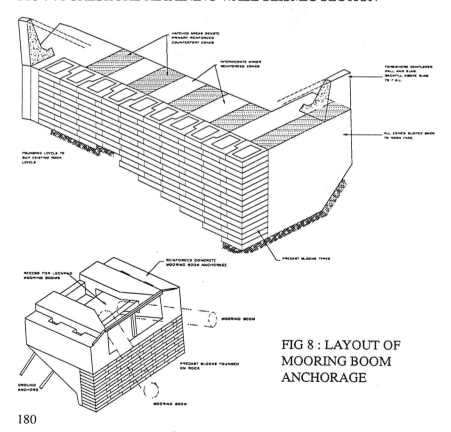

FIG 8 : LAYOUT OF MOORING BOOM ANCHORAGE